Energy and Conflict
in Central Asia and the Caucasus

Energy and Conflict
in Central Asia and the Caucasus

EDITED BY
ROBERT EBEL AND RAJAN MENON

ROWMAN & LITTLEFIELD PUBLISHERS, INC.
Lanham • Boulder • New York • Oxford

ROMAN & LITTLEFIELD PUBLISHERS, INC.

Published in the United States of America
by Rowman & Littlefield Publishers, Inc.
4720 Boston Way, Lanham, Maryland 20706
http://www.rowmanlittlefield.com

12 Hid's Copse Road
Cumnor Hill, Oxford OX2 9JJ, England

British Library Cataloguing in Publication Information Available

Library of Congress Cataloging-in-Publication Data

Energy and conflict in Central Asia and the Caucasus / edited by Robert Ebel and
  Rajan Menon.
    p. cm.
  Includes index.
  ISBN 0-7425-0062-4 (alk. paper)—ISBN 0-7425-0063-2 (pbk. : alk. paper)
    1. Petroleum industry and trade—Caspian Sea Region. 2. Petroleum industry
  and trade—Political aspects—Caspian Sea Region. 3. Gas industry—Caspian Sea
  Region. 4. Gas industry—Political aspects—Caspian Sea Region. I. Ebel, Robert E.
  II. Menon, Rajan, 1953-

  HD9576.C372 E53 2000
  333.8'23'09475–dc21   00-040302

Printed in the United States of America

∞ ™ The paper used in this publication meets the minimum requirements of American National Standard for Information Sciences—Permanence of Paper for Printed
Library Materials, ANSI/NISO Z39.48-1992.

Contents

v

Acknowledgments

This book emerged from a year-long project developed by The National Bureau of Asian Research (NBR) on the relationships among energy, development, and conflict in the Caspian region.

The chapters in this volume have been extensively revised based on a series of interactive research meetings and an e-mail discussion forum involving academics, businesspeople, and policymakers with expertise on the energy-conflict nexus in the Caspian Sea Basin. Earlier versions of the chapters by David Hoffman, Pauline Jones Luong, and Nancy Lubin appeared in the August 1999 issue of the *NBR Analysis.*

The editors thank the authors for their contributions to the project and for their willingness to revise their chapters. In addition, we thank James Clay Moltz, Michael Eisenstadt, Laurent Ruseckas, Elkhan Nuriyev, Mark Brzezinski, S. Enders Wimbush, Dastan Eleukhenov, Jahangir Amuzegar, James Placke, Bulent Aliriza, and Dolkun Kamberi for their comments on earlier drafts of the chapters. We would also like to thank Willy Olsen of Statoil and Michael Townshend of BP-Amoco for speaking at a special luncheon panel held during the final project meeting on April 21, 1999, at the United States Institute of Peace in Washington, D.C.

We are indebted to a number of people at NBR. Without them, the project would have remained an idea and the production of a book would have been impossible. We thank Richard Ellings, NBR President, and Herbert Ellison, former director of NBR's Eurasia Policy Studies program, for their support of the energy and conflict project; Bruce Acker, who helped to formulate and administer the project and also edited the volume; and Erica Johnson, who worked diligently with the editors and contributors to coordinate the preparation of the manuscript and ensure its timely publication. At various stages, other NBR staff lent valuable assistance, particularly Mark Frazier, Jen Linder, Meira Meek, Colleen Altstock, and Marco Milanese.

We are also deeply appreciative of Susan McEachern and her colleagues at

Rowman and Littlefield Publishers for their patience while we went through drafts and solved the typical problems associated with producing an edited volume.

We gratefully acknowledge the United States Institute of Peace (USIP), the Center for Nonproliferation Studies at the Monterey Institute for International Studies, and the Henry M. Jackson Foundation for their generous financial support of the project. In addition, we thank USIP for hosting the April 1999 meeting and the USIP staff, especially Jodi Koviach, who handled our planned, and last-minute, needs with grace and efficiency.

Robert Ebel, Washington, D.C.
Rajan Menon, New York, NY

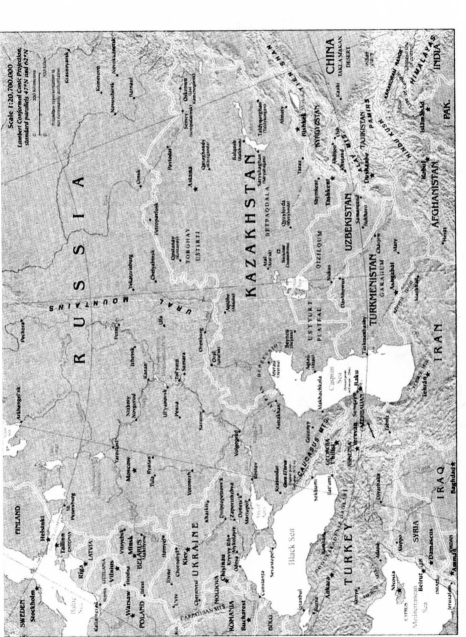

Commonwealth of Independent States

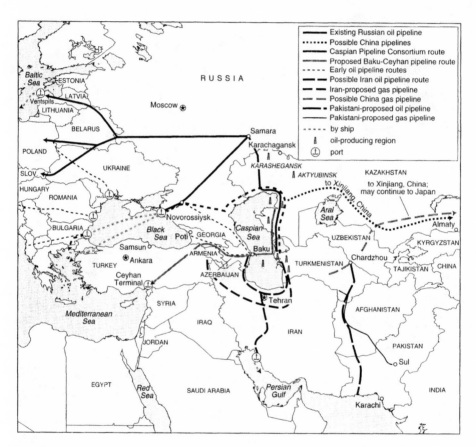

Existing and Potential Oil and Gas Export Routes from the Caspian Basin

1

Introduction: Energy, Conflict, and Development in the Caspian Sea Region

Robert Ebel and Rajan Menon

With the dissolution of the Soviet Union in December 1991, the Caspian Sea Basin was opened to the outside world after having been contained within the Tsarist and Soviet imperial orders for 150 years.[1] Despite the end of the empire, for historical and geographical reasons, Russia still aspires to a dominant role and regards these areas as ones in which it has overriding interests. Yet Russia must now compete for influence in both regions. Moreover, in view of its multifarious problems, it must do so from a position of weakness. The states of the Caspian have attracted attention from corporations and states (and the accompanying political, cultural, and social influences) from the world beyond, not least because of their oil and gas wealth. They have entered the world—and the world has entered them, with consequences that can now only be dimly foreseen.

DOWNSIZING THE CASPIAN

Because of the region's abundant oil and gas reserves, there has been a veritable Caspian mania. Not infrequently, the Caspian Sea zone is described as the world's new great energy frontier and as a place of great importance to world politics in general and to U.S. interests in particular. In part, this hyperbole thrives on simple ignorance: most Americans know next to nothing about this part of the world, and exaggerations go unchallenged—or, more likely, are simply unnoticed. The breathlessness occasioned by the Caspian also stems from the parochial interests of oil companies and academic specialists on the region: the former's wealth and the latter's status are bound up with the level of attention the region gets.

The western press has also given Caspian oil and natural gas far more exposure than is warranted. Unsubstantiated reserve levels are quoted frequently, with the Caspian's oil wealth often compared to that of the Persian Gulf. More realistic assessments exist, but tend to be overlooked, and developments in other parts of the world that are no less significant get short shrift. Consider a case in point. If the AIOC (the Baku-based Azerbaijan International Operating Company, the earliest oil consortium to establish itself in the Caspian after the dissolution of the USSR) is successful, and the pipelines required to handle peak capacity are in place, its fields may produce about 1 million barrels of oil per day (b/d) by 2007—an increase from about 97,000 b/d in 1999. This would certainly be a significant achievement, yet the plans of Petrobras (Brazil) for a comparable increase in production in half the time draws scant coverage. The strategic importance of the Caspian region is likewise overblown: it is asserted routinely that the United States has major interests there. To support this dubious proposition, the area is depicted variously as a latter-day Silk Road, a new arena for great power competition, a venue for battles against Islamic fundamentalism, and the place to block the revival of Russian imperialism. The exuberance sometimes gives rise to proposals that are, to say the least, unwise: to wit, the recommendation for an alignment between the United States and Uzbekistan, a politically repressive, economically troubled, and potentially unstable country whose regional ambitions frighten its neighbors.

RHETORIC VERSUS REALITY

The economic and strategic facts about the Caspian are different, and a reality check is therefore in order. When considered as a proportion of global production or reserves, the Caspian's contribution is likely to be less than 3 percent in 2010. The exaggeration of the region's energy wealth often derives from a conflation of "proven" reserves (recoverable given current technology and energy prices) and "possible" reserves (deposits that are deemed to exist based on computer models or that could be exploited given changes in energy prices and technology).

The Caspian region will be important to global energy, but no more so than, say, the North Sea. As for the United States, a very small portion of its oil imports will come from the Caspian: increasingly most of the oil imported by the United States will originate from the Americas. And many grandiose Caspian pipeline projects will remain pipedreams because the levels of production and the global energy prices required to make them profitable propositions are quite unlikely to materialize, and the political obstacles that thwart them will not soon disappear.

The future of Caspian energy is hostage to many unknowns. Much of the Caspian Sea remains unexplored, and while there are many sites that could contain oil and gas, the sheer statistical probability of large finds in areas that have yet to be explored are low as a general rule, even with the dramatic improvements in prospecting technology. Furthermore, there needs to be a great deal of investment (in drilling rigs, pipelines, pumping stations, etc.) to drill for hydrocarbons and to transport them to distant markets from what is a landlocked region. Only three drilling rigs are operating in the Caspian at present—a major constraint on increases in future production. World oil prices will determine whether the investment needed to cope with these problems materializes. Sooner or later a prolonged downward trend could make the Caspian unattractive to investors, particularly as the costs of extracting oil and transporting it from this region is high relative to other oil-producing areas. There are several conditions (some already at play) that could, in combination, make for persistently low oil prices—a possibility that should not be obscured by the steep rise in prices that began in the latter half of 1999. Among these conditions are the increasing energy efficiency in the manufacturing sectors of the major consuming countries; the development of transportation technologies that may eventually reduce sharply the need for gasoline; the discovery of new and easily accessible reserves that, to boot, stay outside the OPEC cartel; and the ending of sanctions on Iran and Iraq, which would permit these countries, aided by foreign investment, to increase oil production and exports.

As for strategic salience, the Caspian region is hardly comparable to areas in which the United States has longstanding interests and a substantial presence: the Middle East, Western Europe, and Northeast Asia. The United States will never risk much treasure or blood for strategic pursuits in the Caspian. Nor should it: there are no stakes that justify such sacrifices and the requisite domestic political support would therefore be lacking. Nevertheless, the assertion that the Caspian Basin has emerged as a place of great strategic significance is commonplace. But it begs the question of why and how the United States should increase its presence in the region. Any such increase can only be achieved in one of two ways: by reducing commitments in areas where the United States has long had critical interests (Europe and Northeast Asia), or by raising additional revenue that enables a larger presence in the Caspian without diminishing it elsewhere. Balance of power calculations will rule out the first choice, the lack of public support the second. That being the case, the United States risks promoting instability by encouraging states to take dangerous actions that are based on an anticipation of U.S. support that proves false. This is why depicting American stakes in a capacious manner is reckless. In this volatile part of the world, in which the United States has no overriding security interests and which is chock full of ethnic conflicts, contested borders, and shifting power balances, the wiser course is to walk softly and carry a small stick. There are many useful ways in which the United States can, and should, remain engaged—politically and economically—in the Caspian, but they do not require engaging in rhetoric that is detached from reality.

SOURCES OF CHANGE AND THE ROLE OF ENERGY

This intellectual downsizing is not meant to detract from the significance of the Caspian Sea Basin. (That would be a curious thing to do in the introduction to a book devoted to the region.) As readers of this volume will see, fundamental changes are occurring along the Caspian littoral, where, as in the rest of the post-Soviet world, a grand historical experiment is underway. Because of the Soviet collapse, the Russian predicament, and the slow but steady exposure to the currents of what has come to be known as globalization, the Caspian region's future is up for grabs. The political, economic, cultural, and strategic trajectory of the area, which seemed certain for so long, is now entirely uncertain—a work in progress, and in the very early stages at that. Here, as in the other states that emerged from the Soviet Union, newly independent countries are trying to complete, simultaneously and in short order, a prolonged process that, in the West, was achieved more or less sequentially: the consolidation of sovereignty, the formation of national identity, the construction of stable polities, and the creation of market economies.

To make matters worse, the Caspian states are taking on this inherently difficult challenge after emerging from a Soviet experience that has left them woefully ill-prepared. And they are doing so in the context of globalization, which creates opportunities, to be sure, but also makes for myriad vulnerabilities. The ways in which the Caspian's oil and gas are exploited, transported, and used will influence the outcome of this complex experiment. The magnitude of change underway and the role of energy in the outcome make the study of the Caspian region worthwhile and necessary and are the justifications for this volume.

Western geologists had been aware for some time of the rich crude oil and natural gas potential of the Caucasus and Central Asia. Although this also had been known to Soviet authorities, available investment had been directed elsewhere, to the vast fields of Western Siberia, where, it was thought, development would be cheaper and quicker. Natural gas from Turkmenistan and Uzbekistan did flow northward by pipeline into Russia in large amounts. But the oil resources of the Caspian Sea remained untested, while development of the Tengiz oil field in Kazakhstan, the largest oil discovery in the past 20 years or so, remained on hold because the technical challenge of field development exceeded Soviet capabilities. Azerbaijan's oil fields, which had led the world at the turn of the century, were in steady decline with no relief in sight.

Moscow had supplied the outlying republics with all the fuel they required from Siberia—at heavily subsidized prices—through a network of long-distance, large-diameter pipelines. Ironically, the Soviet government's decision, taken years ago, to leave Azerbaijani and Central Asian energy in the ground now threatens Russia's pre-eminence in the Caspian. Pipelines bound Caspian energy producers to Mother Russia. But pipelines may now ensure separation, as the development of new oil and gas fields has led to the consideration of transportation routes—south and west across the Caspian Sea, through Georgia, and through Turkey via Georgia—that circumvent Russia.

Western oil company representatives had traveled first to Kazakhstan, not the better known Azerbaijan, in search of investment opportunities even before the break-up of the Soviet Union. The Soviet oil industry was in decline, badly in need of capital and western technology, and apparently amenable to joint ventures with foreign partners. These discussions were the background to an agreement between Chevron and Kazakhstan to develop the Tengiz oil field as the Soviet Union was falling apart. In turn, President Heydar Aliyev of Azerbaijan moved with determination to entice foreign oil companies to develop the Azerbaijani sector of the Caspian Sea, adding momentum to a process begun by his predecessor, Abulfaz Elchibey. Aliyev's initial success and perhaps his greatest—the "contract of the century," as it would be described—was the agreement between Azerbaijan and a western consortium of oil companies to develop the Chirag, Azeri, and Guneshli oil fields. Other deals quickly followed, for a total of 19 agreements, placing Azerbaijan well ahead in the race for investor commitments. A dispute over the legal status of the Caspian prompted much discussion about whether foreign investment would be inhibited by the resulting uncertainty. But this did not happen even though the littoral states of the Caspian Sea have yet to reach a settlement on its status. Russia, which once harped on the legal dispute and implied that the Caspian's energy resources could not be developed without an agreement, soon decided that its own companies had better get involved in the emerging production-sharing agreements or risk being shut out.

The U.S. government has long encouraged American oil companies to search for and develop sources of oil overseas but away from the Persian Gulf, which continues to be the fulcrum of the global oil market. The reasoning behind this has been threefold. First, while many unexplored areas in the United States might contain a great deal of oil, they will stay off limits for environmental reasons. Second, and consequently, the growing domestic demand for oil will be met only by increasing reliance on foreign oil; the question is where it will come from. Third, if oil imports are to be dependable and energy security is to be maintained, there must be multiple sources of supply. The search for new sources of crude oil in the South Caucasus and Central Asia was consistent with all three calculations. Investment in the Caspian energy sector is, in Washington's eyes, also important to help Azerbaijan,

Kazakhstan, and Turkmenistan consolidate their independence, reduce their extreme dependence on Russia, and develop their economies. Stability and democracy in these countries would be promoted as a result—or so the reasoning went.

The process of finding energy in the Caspian, extracting it, and transporting it to markets from the landlocked region is laden with promise and pitfalls. Oil companies are lured by the prospects of profits and by hopes of developing reserves in a new frontier; but their very unfamiliarity with this region, which was long sealed off by Moscow, makes for uncertainties. The Caspian Sea states, for their part, seek to unlock their oil and gas wealth, which promises secure independence and lasting prosperity—at least in theory. In practice, the challenges of converting hydrocarbon revenue into sustainable growth along with political stability are numerous and vexing. Consider some examples. How can income from energy exports be invested so that the benefits flow to many and not just a few? How can the wealth be spread so that ethnic tensions are not exacerbated in what are, to varying degrees, multiethnic societies? How can new, fragile polities meet the rising expectations of citizens who are unlikely to remain quiescent as in the Soviet era and who, because governments have placed so much hope in energy wealth, may believe that oil and gas are a panacea for their penury? How can the dependence on Russia be overcome without provoking anxiety and anger from a state that is proximate and that may one day overcome its problems and once again become powerful? How can alignments be formed with states that are eager to enter the Caspian zone while at the same time avoiding the dangers of exchanging old dependencies for new ones and of getting caught in the cross fire of great power rivalries? These are just some of the problems on the horizon.

RUSSIAN ANXIETIES: IS A MULTILATERAL REGIME THE SOLUTION?

Neither American nor Russian interests will be advanced if the Caspian region is convulsed by conflict: competition that promotes instability will prove ruinous. For Russia, which faces a plethora of problems at home, turmoil on its southern flank will complicate the pursuit of economic reform and democracy while also reducing Russian security. And western investors in the Caucasus and Central Asia, who desire political stability and the rule of law, will find that there can be no reliable profits absent predictability. For them, the complicated tasks of attracting foreign capital into the Caspian, searching for energy, and transporting it to consumers will be made much harder, if not impossible, by conflagrations. This suggests that there is a shared interest in stability and cooperation.

Yet in Moscow the belief gains ground that, at a time of Russian weakness, the United States, in collusion with Turkey and western oil companies, seeks to displace Russia from the Caspian by controlling the region's energy wealth and pipeline routes. Americans may dismiss such fears as paranoia. But the perceptions are strong in Russia that this is precisely the American plan and that it is being pursued not just in the sphere of energy but through such measures as NATO's Partnership for Peace, which has extended itself into the Caspian region. Perceptions, not just realities, shape policies, particularly in Russia, where anxiety about the country's future as a great power runs high. The question, therefore, is whether Russia, in an effort to prevent its eclipse in a region that is central to its national interests, will pursue policies that promote instability. Russia's power is shrinking; but it is hardly a spent force in the Caspian. Moscow has many ways to exert influence: it is nearby; it is familiar with the region; and it can exploit myriad actual and potential conflicts in Armenia, Azerbaijan, Georgia, Kazakhstan, and Turkmenistan.

In an effort to resist being marginalized in the South Caucasus and Central Asia, Russia could act in ways that sow instability. Stability in the Caspian can most likely be assured through a multilateral arrangement that enables the development of energy in ways that take account of the needs and fears of all players, including Russia. The calculation behind such a system would be that each state with interests in the region would avoid disruptive actions because the gains of cooperation outweigh the losses. Yet discussions about a cooperative regime for the Caspian notwithstanding, movement from talk to action has been conspicuously absent—and there is no sign that this pattern will change. Nevertheless, the principal requirements for any multilateral arrangement can be specified. States with the capacity to act as spoilers must be coopted by bringing them into production-sharing agreements and construction projects so that cooperation brings benefits and disruptive behavior entails tangible losses. Powerful domestic interests must benefit directly from inclusion or see that their broader interests are served by participation. Since upheaval within and among states harms all facets of energy development, a multilateral system must do more than create incentives for cooperation by offering commensurate and tangible gains. It must also provide for institutions with the capacity for early warning, mediation, conflict resolution, and peacekeeping.

Because there are many problems that exist or are waiting in the wings in the Caspian region, an approach that combines co-optation, cooperation, and conflict resolution—and thus restrains unilateralism—is needed. There are festering ethnic conflicts in the Georgian regions of Abkhazia and North Ossetia. The dispute over Nagorno-Karabakh between Armenia and Azerbaijan remains red hot. The polities of Turkmenistan and Azerbaijan—who have their own dispute over the demarcation of the Caspian Sea—are underinstitutionalized and over-personalized, and the future that awaits when Presidents Saparmurat Niyazov and Heydar Aliyev, who dominate politics in their respective countries, pass from the scene is uncertain. True, some anticipated problems have been averted. Relations between Kazakhs and Russians in Kazakhstan have been handled well so far—happily, this particular dog has not barked—but, given the ethnic composition of Kazakhstan, the reality is that every major issue of public policy will perforce involve the nationalities question. This means that interethnic harmony in Kazakhstan is less a matter of finding a permanent solution than it is a process of managing and resolving emerging problems. Likewise, despite periodic breaches, the cease-fire between Armenia and Azerbaijan has held since 1994. But it is too much to expect that the status quo can be frozen and peace guaranteed, because one of the parties, Azerbaijan, rejects the status quo.

But the problems in the Caspian region are numerous and much could go wrong in areas and in ways that cannot be predicted. Given the sheer magnitude of change unfolding in the Caspian zone, the environment in which oil and gas are found and exploited, pipelines are built, and energy revenue is used to pursue economic development could change for the worse. This would affect both states with large energy deposits (Azerbaijan, Kazakhstan, and Turkmenistan) and those that serve (Georgia) or could serve (Armenia) as conduits to export the energy. The turmoil in Chechnya makes this link between regional stability and energy development and transportation clear by raising serious questions about the dependability of the pipeline that takes oil from Baku to the Russian Black Sea port of Novorossiisk via Chechnya. The upheaval in Chechnya is a particular case of a general problem that affects all those involved in the quest for Caspian energy (whether individuals, companies, or states): the inseparability between politics and economics. And it makes clear the need for a multilateral regime that would harness the cooperative potential among players while tamping down their temptation to seek unilateral advantage, but a host of contentious issues that divide the region work to prevent necessity from becoming reality.

PIPELINES: STRATEGY, SECURITY, AND DEVELOPMENT

Washington's support for an oil export pipeline from Baku to the Turkish Mediterranean port of Ceyhan is one of the ventures that the Kremlin sees as emblematic of a policy of undercutting Russia. Ever since the conception of this project—which supposedly had its origins in a January 1995 speech by Marc Grossman, then Assistant Secretary of State—it has enjoyed high-level backing from the U.S. government and has become the centerpiece of relations between Turkey and the United States. In Turkey, it has become a litmus test of American friendship: at this point, AIOC failure to construct the Baku-Ceyhan pipeline could damage U.S.—Turkish relations because Ankara believes that Washington can make the project happen. With subtle reminders to this effect, and by underscoring the environmental dangers of shipping more Caspian oil through the Bosporus and the Dardanelles, Turkey has kept the Baku-Ceyhan project alive at times when it seemed at death's door.

Ankara and Washington insist that Baku-Ceyhan will bolster the political and economic independence of the South Caucasus and Central Asian states. From Moscow's perspective, however, using Baku-Ceyhan to channel Caspian oil west and not north represents an effort by the West to erode Russia's influence in the region. That Washington and Ankara are committed to Baku-Ceyhan despite serious questions about its commercial viability makes Russia all the more convinced that an anti-Russian strategy, not economic logic, drives the project.

The economic rationale for the Baku-Ceyhan pipeline has indeed been questioned. The nub of the issue is whether the cost of building the 1,080-mile pipeline—the estimates have ranged between $2 and $4 billion—will be recouped by "throughput" sufficient to guarantee its financial viability once it is operational. The AIOC's concern has long been that the crude oil reserves of Azeri, Chirag, and Guneshli are currently measured at four billion barrels, whereas six billion barrels are needed to justify construction of the pipeline. While there are other sources of Caspian oil, critics of Baku-Ceyhan point out that those are designated for other pipelines, existing or emerging. The Baku-Ceyhan pipeline will need a million barrels of oil flowing through it daily to be commercially viable, and this may prove impossible from Azerbaijani production alone. The fate of the Baku-Ceyhan line may therefore depend on what Kazakhstan's recoverable reserves turn out to be and on the proportion of Kazakhstani production that could be earmarked for export via Ceyhan. Some experts believe that the onshore Tengiz field could turn out to have between 9 and 13 billion barrels of recoverable oil—far more than initial estimates. Likewise, Kazakhstan's Kashagan offshore field could prove to have several times more oil than even Tengiz; or it could turn out to be a gas condensate deposit similar to Karachaganak in southwestern Kazakhstan. If the optimistic expectations about Tengiz and Kashagan prove justified (and that will not be known until drilling occurs), then some Kazakhstani crude oil might be available for the Baku-Ceyhan project. That, in turn, would make its construction feasible, especially in light of Turkey's commitment to cover cost overruns in excess of $2.7 billion. However, the CPC (Caspian Pipeline Consortium) pipeline from Tengiz to the Black Sea port of Novorossiisk, now under construction, will have a maximum carrying capacity of 1.34 million b/d, making it unlikely that substantial amounts of Kazakhstani crude oil will be made available for export through Baku-Ceyhan in the near future. As for the Turkmen sector of the Caspian Sea, it is basically untested. Any oil discovered there would come too late to help justify the rapid construction of Baku-Ceyhan.

Future Caspian crude production that increases, but falls short of the level needed to make Baku-Ceyhan viable, could be exported by expanding the existing pipeline running from Baku to the Georgian port of Supsa. (That pipeline became operational in February 1999 and has a current capacity of about 120,000 barrels per day.) Likewise, the Baku-Novorossiisk pipeline could be expanded, but only after the struggle over Chechnya has concluded on terms that create durable stability. If these two pipelines are expanded, and there are no major unanticipated increases in production from Caspian fields, the economic basis for Baku-Ceyhan could become shakier still.

Despite concerns about the Baku-Ceyhan project, the persistence of Washington and Ankara—and Turkey's willingness to absorb costs of construction that exceed $2.7 billion—has kept it alive. At the November 1999 meeting of the Organization for Security and Cooperation in Europe (OSCE) in Istanbul, an agreement to build the pipeline was signed by Azerbaijan, Georgia, Kazakhstan, and Turkey. But the attending press coverage notwithstanding, the project may still fail to be completed for reasons having to do with construction costs, the level of Caspian production, and the price of oil.

The future of Caspian energy-producing states depends on export pipelines that are financially viable and secure. Azerbaijan, Kazakhstan, and Turkmenistan are landlocked. Turkmenistan has the added disadvantage of being dependent on Russian goodwill—or, more specifically, on access to Gazprom's pipeline network—to export its natural gas to hard currency markets. If these countries are to develop their economies by shipping energy to the global market, and to do so in a manner that reduces their dependence on Russia, new pipelines are a *sine qua non.* Yet it is obviously easier to note this problem than to fix it. Pipelines are expensive to build and maintain and adequate throughput commitments by oil producers are necessary if pipelines are to move beyond the drawing board. (The requirement for financing natural gas pipelines differs: "offtake" agreements—that is, commitments by buyers to take delivery of an agreed volume of natural gas over an agreed number of years—sufficient to justify construction must be in place.) Because domestic demand for oil and natural gas in the Caucasus and in Central Asia are comparatively small, and likely to stay that way, most of the expansion in Caspian production will be available for export—indeed must be exported. But there can be no expansion in production unless and until export pipelines are in place. The routing of the prospective pipelines involves high stakes. Countries that lie between producers and the ports from which that energy could be exported want to be selected as routes: there are several gains to be made by becoming a channel for energy exports. The transit fees earned can be substantial. Pipeline construction and maintenance generate jobs and have numerous beneficial multiplier effects. The investments made for pipeline construction may bring other sorts of foreign investment in their wake. The transit country's economic and strategic leverage is enhanced. It receives more attention from great powers because its political stability and economic health matter more once pipelines traverse its territory, and it can therefore ask for economic and military assistance from a position of advantage.

The Baku-Novorossiisk pipeline has been shut down since June 1999, and substitute export routes from the Caspian are limited in number and capacity. The Baku-Supsa pipeline is operating at full capacity. Some volumes move by pipeline from Baku to the Caspian Sea port of Makhachkala in Dagestan, where they are then transferred to rail tank cars for delivery to Tikhoretsk, from which point the oil is piped to Novorossiisk. Limited amounts are moved by rail from Baku to Batumi, also on the Black Sea.

Providing the current construction schedule is met—and it may not be, given the delays that have plagued this project—output from the Tengiz field in Kazakhstan will flow through the CPC pipeline via Kazakhstan and Russia to Novorossiisk by the latter part of 2001. Other pipelines are also planned. The Baku-Ceyhan project is one; the pipeline from Kazakhstan's Aktiubinsk and Uzen fields—where China has made significant investments—to China is another. But no other outlets are commercially or politically feasible at present. Even the Aktiubinsk and Uzen fields are too small to justify a long and costly pipeline at this time.

Iran could become an important corridor for the export of Caspian oil to Asian markets, where future demand will be high, especially given the projected needs of China and India. But at present the United States would block the construction of an export pipeline across Iran. While this may change given the shifting political currents in Iranian politics, no dramatic and rapid transformation is likely. Swaps between Iran and Kazakhstan, under which Kazakhstani oil is consumed in northern Iran and an equivalent amount is exported from Iran's Persian Gulf ports and credited to Kazakhstan, have been conducted. But they are not a means to obtain what Iran really wants, which is to become a major outlet for Caspian energy. Tehran and Washington will have to mend fences before Iran can become a major export route for Caspian oil. While a rapprochement may occur, it is hardly something on which Caspian energy producers can stake their development plans.

There are even fewer outlets for Caspian natural gas. Limited volumes are being sent to Iran from Turkmenistan, but the plan for a major pipeline from Turkmenistan through Iran is blocked by the Iran-Libya Sanctions Act of the United States. Nor is it self-evident that Iran, with its potential to become a major gas exporter to Turkey, has an interest in promoting Turkmenistan as a competitor. Russia (which itself buys natural gas from Turkmenistan) prefers that Turkmenistan export to former Soviet republics, thus enabling the sale of Russian gas in hard currency markets. Unfortunately for Turkmenistan, these customers failed to meet their financial obligations and gas deliveries have been halted.

From Washington's vantage point, the best way for Turkmenistan to liberate itself from this crippling dependence on Gazprom's pipeline network is the trans-Caspian gas pipeline. Although the project has received much attention, a long underwater pipeline is expensive for obvious reasons. It also now faces competition from the Shah Deniz gas field in the Azerbaijani sector of the Caspian Sea, which is closer to the Turkish market. Moreover, because the pipeline would traverse Azerbaijan after emerging from the Caspian Sea, its construction could be delayed, even rendered impossible, if Azerbaijan and Turkmenistan fail to settle the dispute over the ownership of the Kyapaz/Serdar field. The trans-Caspian pipeline therefore could very well remain a concept, albeit an alluring one.

The same applies to the plan to export Turkmenistan's natural gas through Afghanistan to Pakistan and on to India. The continuing civil war in Afghanistan makes that a vision that may never become reality. Concepts and vision are nice, but they do little to address Turkmenistan's pressing need, which is to earn money by getting its gas to market. The Turkmen hope of becoming a major supplier to the Turkish market could also be dashed if the anticipated growth for natural gas in Turkey fails to materialize. Another danger is that other suppliers could limit the volume of Turkmenistan's exports. Russia, for instance, could become a major competitor for the Turkish market if the so-called Blue Stream project—designed to deliver Russian gas via a pipeline laid on the bottom of the Black Sea—comes to fruition. There is some relief in sight for Turkmenistan amidst these uncertainties: during the year 2000, Gazprom has contracted to purchase 20 billion cubic meters (bcm) of Turkmen natural gas so that the Russian company can meet its supply obligations. (Payment will be 40 percent in cash and the rest in barter.)

ENERGY AND DEVELOPMENT: PANACEA OR PERIL?

The announcement of deals between Caspian oil and gas states and multinational energy companies and the quotidian maneuverings that serve as the backdrop receive considerable attention. Much less attention is paid to a more fundamental matter: the relationship between hydrocarbon wealth on the one hand and the economic and political development and external orientation of the Caspian Sea states on the other. That is an unfortunate omission, particularly because there is much to be learned from the lessons of other energy-producing countries, particularly those that have relied almost exclusively on energy revenues to promote their economic development.

Azerbaijan, Kazakhstan, and Turkmenistan will be sorely tested once the income derived from oil and gas exports arrives in full measure. Prosperity is a boon, so it would seem, but if the experience of other hydrocarbon-rich states is any guide, managing the wealth generated by oil and gas will present challenges unlike those that the countries of the Caspian littoral have faced so far. The course of their political and economic development, their relationships with their neighbors, and their place in the international order will be shaped by the ways in which they spend, invest, and distribute energy income. There is unprecedented opportunity here, but many dangers as well.

Norway and Britain are the exception when it comes to the effective use of energy revenue. In most other states, the gush of oil and gas wealth has had many malign consequences. The funds have been used to keep entrenched elites in power, to postpone reforms, to underwrite the lavish lifestyles of the privileged, to engage in spending sprees on arms, to build showy and unnecessary projects, and to placate powerful special interests. Among the undesirable consequences have been profligate spending and borrowing, corruption, inequality, repression, and overvalued exchange rates that retard the non-energy sectors of economies. Energy revenues have not been used to build the foundations for economies with diverse types of production and sources of income that, as a result, can achieve balanced economic development. These mistakes need not be repeated in the Caspian, but their prominence in other energy-rich states should, at the very least, make the leaders of the Caspian, and those who study the region from the outside, cautious about viewing energy as a cure-all.

The external consequences of mismanaged oil and gas wealth have been equally pernicious. The cornucopia of advanced weapons—readily available in the global arms bazaar—purchased to placate the military, to acquire regional prestige, and to intimidate neighbors has, in the end, increased insecurity by fostering arms races and regional conflicts. The opportunity costs created by the purchase of vast arsenals have widened the gap between rich and poor and, in so doing, fostered internal instability. (As the fate of the Shah of Iran showed, when economic and social problems mount, an arsenal jam-packed with modern weapons purchased with oil wealth cannot stave off disaster.)

The danger that the misuse of oil and gas revenue could increase tension and conflict within and among states is especially pronounced in Central Asia and the South Caucasus. Borders are disputed in both regions, newly independent states are leery of shifts in the regional balances of power, ethnic minorities contemplate or actively seek secession, and states from beyond (China, Turkey, Iran, Pakistan, and India) are in sharp competition for influence.

Consider the effect that Azerbaijan's oil wealth could have on its relations with Armenia as an example of how wealth generated by energy sales could promote conflict between states. In one scenario, the shared benefits of transporting oil to global markets could usher in cooperative arrangements between the two foes that reduce the danger of war and prompt bold and creative solutions to the dispute over Nagorno-Karabakh. But there is another possibility: if a solution to the dispute over that contested region remains elusive, Azerbaijan could build up its arsenal using its oil wealth and decide to retake by force what it has been unable to regain by diplomacy. The Karabakh issue has such deep resonance in Azerbaijani politics that no leader can ignore this unfinished business. The risk of external powers—Russia, Iran, and Turkey—being drawn into the resulting conflict would be considerable. The Caspian is no longer part of the Soviet empire in which conflict is managed or suppressed by a robust central authority; it is a region in which the combination of Russian weakness and far-reaching change open up unprecedented opportunities for other states to make gains.

The volatile combination of internal instability and changes in the balance of power among states has been a standard cause of conflict in world politics. In the Caspian region, this pattern could replicate itself if energy revenues make for shifts in national power while failing to address the economic and social problems that plague the lives of people and, in so doing, further breed discontent. Power shifts among states and tensions within them could make for a flammable mixture. Energy revenues are seen as a solution to this problem, not least by Caspian leaders who tend to regard them as the equivalent of a cure-all. But the risks of banking heavily on an energy bonanza are considerable. Much could go awry. Not all wells drilled yield large deposits of energy: some may be dry, others may prove paltry. Iran and Iraq could become available for western investment and begin selling significant amounts of oil and (in Iran's case) gas onto the world market. Wars and upheavals could scare investors away from the Caspian and redirect much of the investment capital that might have gone to the South Caucasus and Central Asia elsewhere. If, for these and other reasons, prosperity cannot be assured, will the people of the Caspian region accept the results without complaint? Or will the failure to meet rising expectations lead to unmanageable discontent, especially once the Soviet generation schooled in and habituated to political quiescence passes from the scene?

Investments in the energy sector and the broader forces of globalization will bring many external influences to bear on the Caspian, a domain of weak economies, fragile polities, and quarreling nationalities. Will the region benefit from globalization such that war and instability recede and prosperity and stability advance? Or will the economic and cultural influences of globalization generate instabilities by creating new economic vulnerabilities and exposing citizens to new ideas that call the existing order into question? How will the outside powers that have entered the Caspian conduct themselves? What possible conflicts and coalitions could emerge among them? Will the result be a stable balance, or pernicious rivalries that breed turmoil?

A PREVIEW OF THE VOLUME

These questions, and the others raised in this introduction, are addressed in the chapters that follow. Michael Mandelbaum sets the stage with an account of how the post–Cold War international system shapes and is shaped by the Caspian Sea zone. Mandelbaum argues that the Caspian region is a terminus for many ideas and influences—some from the past, others more recent. He lays out the major characteristic of the emerging global order, considers the place of the Caspian zone within it, and examines how the region could be shaped by the world beyond. The region, he tells us, is exposed simultaneously to various forces from bygone eras: ideas of liberalism and markets that originated in the eighteenth century; the legacies of nineteenth century imperialism; and the pervasive influence of communism in the twentieth century. Mandelbaum identifies globalization and ethnic conflict as the principal sources for change in the twenty-first century—and the major challenges that will determine the destiny of the Caspian states.

The chapters of Terry Lynn Karl and David Hoffman consider—in complementary ways—how energy revenues could influence Azerbaijan's political and economic order.

Karl examines the prospects for Azerbaijan, Kazakhstan, and (to a lesser degree) Turkmenistan in light of the experiences of other energy producing states. What she finds, on balance, is not encouraging. Despite the clear and abundant lessons to be learned when it comes to the "dos" and "don'ts" of using energy revenue effectively (i.e., promoting just and balanced development), she finds that states tend to repeat the errors of the past—not because of ignorance, but because the short-term political interests of leaders reign supreme. Karl acknowledges positive signs and specifies the policies that the Caspian states and the external donors should adopt to increase the chances of success, but pessimism (or call it realism) pervades her analysis.

Hoffman contends that the government of Azerbaijan has put all of its hopes in the oil sector, not just to develop the country's economy but also to increase its political and strategic connections with the West. He concludes that if oil wealth is to contribute to broad-based development, the circle of political decision-making—now small and containing the president, a few key officials, and a number of patronage networks—will have to be enlarged to give parliament and civil society a role. Hoffman believes that oil wealth will increase the power of the central government over society and reduce the leverage of regional authorities—and that neither trend augurs well for the growth of pluralism and accountability.

The chapters of Pauline Jones Luong and Nancy Lubin explore the link between energy and development by focusing on Kazakhstan and Turkmenistan respectively. Echoing Hoffman's observations on Azerbaijan, Jones Luong argues that the government of Kazakhstan has staked its entire development agenda on oil and gas revenues and has raised popular expectations in the process. The government's plan seems to be to acquire popular legitimacy and acquiescence by using energy revenues to raise living standards. But Jones Luong is decidedly pessimistic about this strategy. She concludes that it rests upon unrealistic expectations about the magnitude of possible energy revenue: these have already led to excessive borrowing, unproductive spending, and a neglect of other economic sectors. In contrast to Hoffman's conclusion regarding Azerbaijan, in Kazakhstan Jones Luong sees the danger of public protests spurred by unmet expectations, regional economic imbalances, growing corruption, and the hardships created by the neglect of the non-energy sectors.

Lubin's prognosis for Turkmenistan is, likewise, pessimistic. Relative to the other Caspian energy producers, Turkmenistan has only a small amount of oil (1.2 billion barrels). Its wealth lies in natural gas, of which it has reserves of 101 trillion cubic feet (2.8 trillion cubic meters), the eleventh largest in the world and accounting for 1.9 percent of world reserves. The key issue for Lubin is whether substantial gas revenues will be realized and used effectively before popular disaffection grows to dangerous levels. This depends, in her view, on whether alternatives can be found to Turkmenistan's reliance on Gazprom's pipelines. The outlook is not promising. Only a small amount of Turkmen gas is being exported to Iran, and economic and political obstacles bedevil the planned trans-Caspian pipeline. In the present, what Lubin observes is "waste, crime, hardship. . ., potential instability" and the growing challenge posed by drug traffickers who are using Turkmenistan as an export route. The weight of these problems is increased by the near-total power wielded by President Niyazov, his frail health, and the uncertainties of how the succession that follows his departure will be managed.

The contributions of Karl, Hoffman, Lubin, and Jones Luong illuminate the links between revenues and a range of domestic issues: corruption, the distribution of wealth, the balance of power between central and provincial governments, and the standing of civil society vis-à-vis the state. Their chapters serve as a reminder that the Caspian should not be studied in isolation—as if it were *sui generis.* Much has been written about the ways in which energy wealth shapes polities, economies, and societies. We know, for example, that hopes that energy will be a cure-all are more often dashed than not and that the range of possibilities is broad, embracing, for instance, Norway, Nigeria, Saudi Arabia, Venezuela, Mexico, Iraq, and Indonesia. Thus the Caspian's present and future should be related to the experiences of a larger universe.

This volume also examines the connection between the prospects for energy development and the politics within and among states in the broader Caspian region. Shireen Hunter's chapter highlights the fear aroused by, and the threat posed to, Central Asian regimes by the radical Islam of Afghanistan's Taliban movement. She shows that the radicalism of the Taliban has prompted cooperation between Tajikistan and Uzbekistan and a limited improvement of relations between Iran and Uzbekistan, even though relations between both pairs have been strained. Turkmenistan by contrast has sought to work with the Taliban, evidently calculating that Turkmen natural gas can be shipped to Pakistan (and perhaps markets beyond) through Afghanistan if the Taliban brings stability to that country. Hunter notes that Turkmenistan is uncertain about the trans-Caspian pipeline and doubtful that Iran will become a major export corridor for Turkmen energy given American opposition. She concludes that these calculations help account for Turkmenistan's favorable orientation toward the Taliban.

Martha Brill Olcott explores the interrelationship between Central Asia and the Caucasus, two proximate areas that are rarely considered in tandem. Olcott detects a nascent sense of regionalism that subsumes the two areas. She suggests that the transportation of Central Asian energy through Azerbaijan and Georgia, and progress in the European Union's TRACECA project ("Transport Corridor Europe Caucasus Central Asia," conceived in 1993 to further east-west trade through Central Asia and the South Caucasus) could make this regionalism robust. Developments in the security realm could add to the momentum for cooperation between Central Asia and the South Caucasus. The most important developments in this direction are the participation of Georgia and Azerbaijan in CENTRASBAT (Central Asian Battalion, a venture among Kazakhstan, Kyrgyzstan, and Uzbekistan) exercises, the activities of NATO's Partnership for Peace, and the GUUAM (Georgia, Uzbekistan, Ukraine, Azerbaijan, Moldova) coalition. It is early in the game, and the institutional foundation for deep and enduring cooperation between Central Asia and the South Caucasus does not yet exist, but the processes Olcott brings to light nevertheless reveal new patterns emerging in the post-Soviet landscape.

Geoffrey Kemp considers the outlook for Iran and the implications for Caspian energy. His analysis appears at a time of considerable uncertainty about Iran and its relationship with the United States. American opposition to Iran's emergence as a conduit for Caspian Sea oil and gas is manifested in the Iran-Libya Sanctions Act. That legislation has prevented Iran's emergence as a corridor for the transportation of energy from the region even though, on purely economic grounds, oil companies see an Iranian route as a better alternative to Baku-Ceyhan. Yet both the companies and the Caspian states are watching the ferment underway in Iran and assessing how it might reshape U.S.–Iranian relations. The battle between the proponents and opponents of change in Iran has also prompted a debate in the United States and in Iran itself about rethinking the bilateral relationship. Kemp argues that Iran's policy toward the Caspian has been pragmatic, not doctrinaire. There has been a convergence of interests with Russia: both countries perceive an American effort to marginalize their role in the export of Caspian energy, and neither wants upheaval in Central Asia and the Caucasus. Nevertheless, Iran and Russia could become competitors in the Caspian: geography makes both countries potential corridors for exporting the region's energy. Iran is already serving as a market for limited amounts of gas from Turkmenistan and has also been engaged in oil swaps with Kazakhstan. The latter arrangement gives Iran a role in Caspian energy despite U.S. opposition. But it also makes sense because most of Iran's population is situated in the north of the country, while its oil and gas reserves (the latter are the second largest in the world) are in the south.

Yet Iran's future role in Caspian energy exports is limited by several conditions. Turkmenistan, encouraged by the United States, is increasingly looking to the trans-Caspian pipeline, not Iran, as its best option. Energy swaps are no substitute for serving as a major corridor for Caspian energy—the latter being an unlikely prospect for Iran in view of U.S. opposition. Moreover, Iran's own energy infrastructure is in disrepair while its economy suffers from numerous problems. Even the easing or lifting of U.S. sanctions would, therefore, not result in rapid increases in Iran's production and exports.

Peter Rutland examines how the independence of the Caspian states and the weakness of Russia have affected and will continue to affect Russia's position in the region. Russia has dominated the Caspian region since the nineteenth century and continues to regard it as one in which it has preeminent, even special, interests. Yet it faces unprecedented challenges, as the ex-Soviet Caspian states seek to move out of its shadow, as states from beyond the region move in to challenge its longstanding hegemony, and as the currents of globalization penetrate the region with consequences for Russia's position that can only be dimly foreseen.

One of Russia's chief competitors is Turkey, which is aided by proximity to the Caspian that is both spatial and cultural. Sabri Sayari analyzes Turkish perspectives and policy toward the region and the nature and limits of its influence. Linguistic and cultural affinity links Turkey and the Muslim states of Central Asia (except Tajikistan, whose affinities are with Iran) and Azerbaijan. Turkey also aspires to play a major role in the two regions, whose states want to diversify their economic and political ties and who, therefore, regard Ankara as a natural partner, even patron. These realities work to Ankara's advantage, but others limit what Turkey can accomplish. Among them are Turkey's Kurdish problem, its economic limitations, and Armenia's strong aversion to the growth of Turkish influence, which makes Erevan tack toward Russia. In all, argues Sayari, Turkey has adopted a pragmatic and flexible policy toward the Caspian region. To meet the rapid growth in its need for imported energy, Turkey signed a major agreement in 1996 to buy gas from Iran. It entered the following year into a joint venture with Gazprom and ENI (of Italy) for the Blue Stream project designed to carry Russian gas to Turkey under the Black Sea. In 1998, Turkey and Turkmenistan signed an agreement on the trans-Caspian pipeline. If built, this expensive project will bring Turkmenistan's gas to Turkey via the Caspian. Sayari argues that Turkey's keen interest in the Baku-Ceyhan pipeline (which it seeks for environmental, economic, and strategic reasons) gives it a stake in a peaceful settlement of the Nagorno-Karabakh dispute and that Ankara has sought, albeit without success, to play a diplomatic role to that end.

China has moved quickly to increase its presence in Central Asia, and Dru Gladney looks at the determinants of its policy. China sees Central Asia as a potential source of oil—which is important to Beijing given that it became a net oil importer in 1993 and seeks to reduce its dependence on the Middle East, to which it is connected by long, vulnerable sealanes. China has moved with particular dispatch into Kazakhstan, where it has acquired oilfields from which it plans to transport energy east to its domestic market. Beijing also realizes that turbulence in Central Asia could affect the integrity of China. Central Asia adjoins Xinjiang, whose Uighur people, like most Central Asians, are Turkic Muslims. In the post-Soviet era, China faces increasing ethno-religious nationalism in Xinjiang and worries that Turkic nationalism and fundamentalist Islam could intrude into its already-turbulent eastern province from Central Asia, which has a Uighur population and is used as a base by Uighur émigré groups. Yet the transportation links that are expanding between China and Central Asia and the consequent growth of trade and tourism are a double-edged sword. They could make for a mutually beneficial interdependence, but also could create apprehensions in Central Asia about Chinese dominance while increasing the flow of weapons, political and religious literature, and separatists from Central Asia into Xinjiang.

NOTE

1. For the purposes of this volume, the Caspian Basin, the Caspian region, or simply the "Caspian" consists of Azerbaijan, Kazakhstan, and Turkmenistan. It is defined by geology, specifically the location of confirmed oil and natural gas deposits, rather than demography. When the problem of moving this oil and gas to market is taken into account this geological definition encompasses an area stretching from China to the Black Sea and from Russia to the Persian Gulf. As Laurent Ruseckas notes, the conceptual definition of the Caspian is driven by oil and replaces the map based on the Soviet era. See Laurent Ruseckas, "State of the Field Report: Energy and Politics in Central Asia and the Caucasus," *AccessAsia Review* 1, no. 2 (The National Bureau of Asian Research): 43.

2

The Caspian Region
in the Twenty-first Century

Michael Mandelbaum

The subject of this volume—the Caspian Sea region—can be defined in straightforward geographical terms. But to understand the present and the future of the region, it is helpful to move into the realm of metaphor. The Caspian region is a crossroads, a meeting point of different people, different cultures, and different political, social, and economic forces. As such, it can be compared to the central railway station of a large empire, into which flow people and products from the empire's various districts. The most striking difference among the metaphorical travelers that come together in this region is chronological. For the region is the site of the convergence—and sometimes the collision—of political and economic ideas and social forces that have their origins in four different centuries: the eighteenth, the nineteenth, the twentieth, and even the twenty-first. The region's future will be determined by the interplay of these time-traveling historical forces.

CURRENTS FROM THE EIGHTEENTH CENTURY

The year 2000 is the two-hundredth anniversary of the end of the eighteenth century. It was a long time ago. Yet, oddly, some of the most powerful currents at work in the Caspian region as the twenty-first century dawns reach back to those days before the advent of electric power, steel, or scientific medicine. It was in the eighteenth century that the most influential political and economic ideas of the present day first found their footing. From the eighteenth century dates political democracy as we know it, which flourished initially in the British Isles and on the eastern coast of North America. It was in the eighteenth century that the first systematic account of free markets was

21

given by Adam Smith. With the end of the Cold War, these are the dominant forms of politics and economics everywhere, including in this region.

"Dominant" here means unchallenged in theory, with the alternatives discredited. Liberal political and economic ideas are not necessarily well established in practice everywhere; and in the Caspian region they are sufficiently distant from the historical experience of the peoples that live there to qualify as alien. But for want of any alternatives, liberal politics is the standard to which all in the region aspire, at least rhetorically, and according to which they will be assessed by outsiders. It has become a convention in the post–Cold War period to judge sovereign states by the extent to which they have in place liberal political institutions and market-oriented economic practices. This convention is embedded in a term that has become a familiar part of post–Cold War discourse: *the transition.*

To be sure, the countries of the Caspian region are liberal in neither political nor economic terms, and the chances that they will reproduce the sovereign parliaments and free markets of the West in the near future are negligible. For this there are two reasons. First, what have emerged in the wake of communism as the necessary bases for genuine democracy and functioning free markets are missing in the region. Democratic politics in most countries are practiced by literate and relatively well-to-do citizens and are underpinned by a thick network of secondary associations known as civil society. Markets require institutions such as courts and banks, and trained personnel—lawyers and accountants—to operate them. The countries of the Caspian region are conspicuously lacking in all of these.

Second, post-communist leaders of the countries of the Caspian, although they pay lip service to democracy and free markets, are not in fact particularly well disposed to either, and will resist, insofar as this is possible, their full establishment. The supreme goal of almost all these heads of government is to extend their own reach and strengthen their own grip on the societies they govern. Democracy and markets diminish their reach and weaken their grip.

Still, the legacy of the eighteenth century is not altogether irrelevant to the region. Liberal ideas and practices do have some resonance there, as they do everywhere. Some halting, partial efforts are underway to put them into practice. Moreover, it is possible to say with only modest ethnocentrism that the hopes of this region for tolerable politics and prosperous economics, all other things being equal, do rest on the extent to which the countries there can move in the direction that the idea of the transition denotes. Finally, the efforts of the region's post-communist rulers to the contrary notwithstanding, the market, at least, will intrude on the societies of the Caspian region. The economic wall the communists built around it has come down and cannot be rebuilt. In these countries, as everywhere else in the world, the workings of the market will bring change, sometimes on a large scale, and often unexpected and unwelcome. This will help shape the region's future.

THE RETURN OF THE "GREAT GAME"?

The legacy of the nineteenth century is also visible in the Caspian region. Sometimes this visibility is something of a mirage, as in the case of what is often thought to be the most prominent carryover from the previous century: the "Great Game." The analogy is frequently made between the scramble for energy riches now and the imperial rivalry then between Russia, coming from the northwest as part of the expansion of the Russian state that began in the sixteenth century, and Great Britain, rising periodically from its bastion on the Asian subcontinent to check the northern bear.

The analogy's appeal is plain, and the parallels striking. The region in question is the same. In both cases, high stakes have been involved—in the nineteenth century the control of territory and the security of what were then the two largest empires on the planet; today the enormous wealth that flows from energy resources. While analogies are useful, however, they can also be misleading. The differences between the Anglo-Russian rivalry in the nineteenth century and the politics and economics of Caspian energy now are more important than the similarities.

One difference is captured by the term itself. "The Great Game" has a romantic overtone. The romanticism associated with this great imperial rivalry was typical of the nineteenth century, or at least of the twentieth century's retrospective view of it. This was a game played by brave, eccentric, often lone adventurers who journeyed into hostile, unknown territory, sometimes dying tragically, or heroically, or both. In Britain in the nineteenth century the names of the players were well known: Conolly, beheaded by the Emir of Bokhara; Burnes, martyred in the first Afghan war; Younghusband, the last in the line of great explorers of Chinese Turkestan. There were also Russian equivalents who were known in Russia in the nineteenth century but whose names were blotted out by a Bolshevik regime that had no use for them. These individuals played the same role in the culture as the great explorers of the nineteenth and early twentieth centuries: Stanley, Sir Richard Burton, Lewis and Clark, the polar explorers Scott and Amundsen. They were especially appealing figures because their histories exalted the individual in an age that was already beginning to be dominated by machines and organizations.

A hundred years later, the exploitation of Caspian region energy is preeminently, quintessentially, the work of powerful machines and vast organizations. The participants are teams of executives, geologists, engineers, and bankers. The contrast with the lone heroes of yesteryear could not be starker.

Where the metaphor of the game is concerned, the differences are also pronounced. The players of today are different than the players of the nineteenth century. Then there were two more-or-less equal contestants, both present in full force in the stretch of Asia for which they were battling. They were like prizefighters glowering at each other from opposite corners of the ring. In today's game of Caspian energy, there are many sovereign states, not just two, and there are many private interests. Russia, of course, is still important for reasons of geography, but the United States cannot substitute for British India because it is not and will not be physically present. American interests are not substantial enough to justify the dispatch of an expeditionary force to the Caspian region. If the politics and economics of the region turn into a military contest, the United States will not wage it directly. But it is unlikely to be a military contest.

This is not to say that harmony is destined to reign in the Caspian region. Great-power rivalry, even if not in its pure nineteenth-century form, is not necessarily entirely irrelevant to its future. It is in fact the site of wariness in the present. It might also conceivably be an arena for conflict (more likely political than military) in the future on the part of countries that qualify, fully or in part, as outside powers: Russia, China, Turkey, Iran, and the United States. All have interests in and preferences for the Caspian region. Those interests and preferences are not identical, and, insofar as they bear on the future of the region, they will not necessarily be expressed cooperatively.

The enduring legacy of the nineteenth century is not imperial rivalry but rather the continuing impact of the region's absorption into the tsarist empire. As a consequence of the imperial experience, the main transportation routes from Central Asia and the Caucasus run through Russia. Expanding and redirecting them will require time and money. Additionally, as a result of the nineteenth-century experience, the region is home to a substantial Russian diaspora. The largest and politically most important one inhabits the northern tier of Kazakhstan, along its border with Russia itself. The status of that diaspora is a potential source of conflict between the two countries.

THE COMMUNIST LEGACY

The dominant influence on the Caspian region in the twentieth century was, of course, communist rule, the end of which marks the starting point for a new period in the history of the region, the period in which it now finds itself. It is tempting to believe that the effects of this most recent period will be the least enduring, that they are already fading and will quickly disappear. On the surface it might seem that this is so. Communist institutions, communist public rhetoric, the total political and economic aspiration of real existing communism, all are now gone. And it is certainly to be hoped that one of the hallmarks of communism, namely coercion and repression of a severity never practiced (indeed perhaps never even contemplated) by the tsars of the nineteenth century and the traditional rulers who preceded them, now belong entirely to the past. But, of course, the effects of the communist experience have not disappeared. Those effects—poisoned air and water and flattened, fragmented societies—will be felt far into the future. The communist experience was and is for the region like a childhood trauma that will shape the experiences of a lifetime.

Moreover, one enduring and important feature of the communist period has not only survived, it has been ratified—it might almost be said sanctified—in the post-communist period: the communist borders. Here, as in the Balkans, internal administrative lines were promoted to the status of sovereign frontiers. But those borders, some of them inherited from the tsarist period, were in many cases arbitrarily drawn and sometimes divided peoples who believed they belonged together. Thus the lines were not always popular, legitimate, or stable. This has contributed to another feature of the twentieth century that preceded, was suppressed by, and now has survived communism to be an all-too-prominent feature of present day Caspian affairs: ethnic and national conflict.

Such conflicts have often arisen over controversies about the location of borders. They concern the distribution of sovereignty, with which peoples became sufficiently dissatisfied to fight to change them. This is what triggered, for example, the Armenian-Azeri conflict over Nagorno-Karabakh. Conflicts in this region and elsewhere have had another familiar, unhappy characteristic: the involuntary movement of peoples in large numbers—now known, thanks to its outbreak in the Balkans in the 1990s, as "ethnic cleansing." Conflict and displacement are very much a feature of the present day Caspian region, having occurred in all three countries of the South Caucasus, and in nearby Tajikistan and Afghanistan. In a more orderly way, and in smaller numbers thus far, Russians are leaving the former Soviet republics in which they found themselves after the dissolution of the Soviet Union.

Another characteristic of twentieth-century international politics relevant to the Caspian region is the systematic application of science and technology to the techniques of warfare. So successful, or perhaps more aptly so potent, was this combination that a new term was coined to denote the results: weapons of mass destruction. The techniques for fabricating weapons of mass destruction are now, in the case of nuclear weapons, a half-century old. While not readily available, they are scarcely impossible to obtain. Three countries with interests in this region do have nuclear weapons: the United States, Russia, and China. One other outside power—Iran—seems bent on obtaining them. Because they are so powerful, to the extent that nineteenth-century-style rivalry becomes a feature of this region these twentieth-century instruments of warfare are sure to be relevant to its affairs.

Newer technologies, however, are expected to dominate not only the military but also, and it is to be hoped more importantly, the economic affairs of the twenty-first century. Will the expected (or at least hoped for) dominant trends of the next century be represented in other than token fashion in a region scarcely on the cutting edge of history? The world is said to be entering the age of globalization. Its main feature is the expansion of commerce as a result of the revolution in communication brought about by new information technology. The consequences of this trend in the most hopeful projection are the supremacy of economics over politics, a corresponding rise of international and internal cooperation at the expense of conflict, and a corollary decline in the power and importance of the state accompanied by the rise in the significance of individuals and non-state actors. The affinity of these trends for the ideas of the eighteenth century is plain, and the combination of the two will lead, it is hoped, if not to the best of all possible worlds then at least to a better twenty-first than twentieth century.

The hopes of the Caspian region to take part in this happy experience rest on the energy resources that it harbors. The energy deposits in the region will not shake the world. They are not large enough. They are dwarfed by the energy of the Persian Gulf. But Caspian region energy will shape the future of this particular region, and, in the best case, lift its countries toward western levels of well being. There are potential connections between energy and globalization. Energy resources certainly give an economic emphasis to the affairs of the region. Cooperation is required to extract, transport, and sell these resources, and they are being, and will be, mainly extracted, transported, and sold not by governments but by private firms—nonstate actors, although not small, helpless ones.

Yet great wealth can be a curse as well as a blessing, and great wealth in the form of subterranean deposits of hydrocarbons, under the control of sovereign states and their rulers, is as often malignant as it is beneficial in its effects. Access to large sources of energy has not always been associated with either western-style politics or the untrammeled market of Adam Smith's ideal. To the contrary, such access has led, in places all too similar to the countries of the Caspian region, to concentrations of wealth and power; exacerbating inequality and subverting political liberty. In such countries oil wealth has been not a platform for, but rather an obstacle to, the establishment of modern, self-sustaining, broad-based economies. It is an irritant to, not a lubricant for, social and economic tensions. Already the questions of jurisdiction in the Caspian Sea and the routes for pipelines from the region have become occasions for conflict rather than cooperation. Perhaps the most important question hanging over the future of the Caspian region, therefore, is what effects its energy wealth will have. And this question is relevant as well to the chronologically diverse forces that will bear on the region's future.

The effect of the convergence of political, economic, and social forces with roots in different centuries on a region long shielded from most of them is impossible to predict. But some sense of the possibilities can be gleaned by returning to our starting point—the realm of metaphor. The Caspian region may turn out to be a kind of museum, with different rooms for the effects of different eras. Some countries may bear the marks of one century; others another. (This may be true of different regions within the same country.) Some parts of public life may be dominated by the legacy of one era while others are marked by legacies of another period.

A second metaphor suggests a different result. The interaction of the relevant social, economic, and political forces in this region may resemble the genetic competition that drives the process of evolution. Some genes may turn out to be dominant and survive, while others prove to be recessive and die out. The likeliest future for the region is captured by yet a third metaphor: the pressure cooker. Like a pressure cooker, the interaction of the various relevant ingredients in the Caspian region's future will produce something different from any of them and not entirely predictable. The result will be flavorful or noxious according to taste. If the metaphor of the pressure cooker offers a glimpse of the likeliest future for the region, it also suggests the worst one. What comes out of a pressure cooker is always a matter of taste, but on one point all cooks agree: the least desirable outcome, the one to be avoided at all costs, is an explosion.

3

Crude Calculations: OPEC Lessons for the Caspian Region

Terry Lynn Karl

The Caspian Basin has been called the next Persian Gulf, an area of such enormous untapped energy resources that it will rival the Middle East. Azerbaijan is touted as "the Kuwait of the Caucasus" and Kazakhstan as "the second Saudi Arabia" by those anxious to spread the word of the region's potential. Rulers in the Caspian Basin promise that the exploitation of their black gold will bring enormous benefits to the region's inhabitants: independence after more than a century of Russian domination, prosperity through rapid economic growth based on market economies, political stability, and even democracy. With some experts proclaiming that the Caspian countries will play a major role in all future planning on world energy supplies, oil euphoria has been rampant. "The region's wells will shower unimaginable wealth on people whose annual per capita GDP today hovers between $400 and $600," one reporter writes, "building a new El Dorado in nations where camels still outnumber automobiles in 1998."[1]

Buttressing these glowing predictions is the fact that Azerbaijan, Kazakhstan, and Turkmenistan have the potential to be among the few "lucky" so-called capital surplus oil exporters. Like Saudi Arabia, Kuwait, Libya, Qatar, and the United Arab Emirates, they have relatively small populations in relationship to their projected amounts of oil and natural gas. In this respect, they appear to be more fortunate than the capital deficient countries of Mexico, Iran, Algeria, Indonesia, Nigeria, Venezuela-Trinidad-Tobago, Ecuador, Gabon, Oman, Egypt, Syria, Cameroon, and others. In addition, they also possess gold, uranium, copper, and other minerals,[2] which means that there should be plenty of petrodollars and other resource-generated wealth to spread around.

The signs of a boom are already present. In Baku, the grand mansions built by Azerbaijan's oil millionaires a century ago now house fancy restaurants and expensive perfumeries, while in Almaty designer boutiques and American sport utility vehicles beckon the people of Kazakhstan. Ashgabat, the desert capital of Turkmenistan, has a brand-new football stadium (despite barely having a team) and its newly paved roads are lined with swanky hotels. Armored Mercedes-Benz cars are seen in the streets of each city, while clubs are jammed with petro-barons and their hangers-on from Russia, Texas, California, Iran, and Turkey. "Stay in town long enough," an Azeri oil executive remarks, "and you will meet every major oil player in the world."[3] All are drawn by the lure of liquid gold: in this last great oil rush of the twentieth century, the stakes are very high.

But will this bonanza actually materialize? And if it does, will it bring the political autonomy, economic development, peace, and liberty that the leaders of these countries promise and their people so desperately need? In the decades to come, will the countries of the Caspian Basin begin to look more like Norway, with its sustainable growth, environmental protection, and vibrant democracy, or will they resemble Nigeria, overwhelmed by rent-seeking, the destruction of its lands, and the rise in ethnic conflict? Will the legacy of this newest scramble for petroleum be a higher standard of living and a more representative polity—or will it merely be a number of black holes left behind long after the oil companies are gone? In short, what will be the impact of this special conjuncture on the oil-exporters of the Caspian Basin?

This chapter argues that, under current conditions, the oil-rich countries of the Caspian Basin are unlikely to be able to transform their mineral wealth into sustainable economic development and political democratization—at least in the medium term. To the contrary, the experience of most OPEC countries suggests that there may be short-term benefits and long-term costs to their resource wealth. While petrodollars may provide a temporary boost to the economy and help to buttress some badly needed political stability, the discovery of huge petroleum reserves may eventually produce a highly skewed, inequitable, and mono-export-dependent model of economic growth, as well as the exacerbation of existing tendencies toward kleptocracy and authoritarian rule. The impact on regional conflict could be especially pernicious: in the short term, the hope is that energy development might reduce regional ethnic tensions in Nagorno-Karabakh, Abkhazia, South Ossetia, North Ossetia, and Chechnya by benefiting diverse actors in these various conflicts. But in the longer run, as these countries grow increasingly dependent on petrodollars to smooth over their problems, any disruption in these revenues from plunging prices or interference with pipelines could well portend a sharp rise in regional conflict.

History, unfortunately, does repeat itself—even when the prescriptions for doing otherwise are abundantly clear. Despite the fact that economists now agree on a general set of policies that might produce economic growth and diversification away from dependence on petroleum, most of these prescriptions are unlikely to be implemented by political leaders or their private sector allies—at least in the form they are intended. Policies that determine the utilization of oil wealth are not determined by long-term economic calculations—a reality that few economists seem to understand. Instead, they are shaped by oil-fueled grandiose aspirations of rulers, the intense politicization of the policy process, and the predatory activities of elites—the central consequences of introducing oil rents into especially weak institutions. Indeed, this is the very essence of the petro-state, and it does not bode well for Caspian development.

BECOMING PETRO-STATES: THE SPECIAL CHALLENGE OF THE CASPIAN STATES

There can be little doubt that the countries of the Caspian region, like other oil exporters, are increasingly defined by the exploitation of their energy resources. The Caspian is a premier area for the export of petroleum due to the projected size of reserves, which are considered important by any standards. The most optimistic observers have even claimed Caspian oil production could reach 6 million barrels per day (b/d); this would be a sevenfold increase over current output and is equivalent to the production of the North Sea. Proven natural gas reserves, estimated at 200–350 trillion cubic feet, are comparable to those of North America, and unproven gas reserves could even double that figure.[4] But even if reserves prove significantly smaller than expected and the most pessimistic projections of the performance of the international oil market prevail,[5] expected windfalls, whatever their eventual scale, will have a dramatic impact on the political economies of Azerbaijan, Kazakhstan, and Turkmenistan. Using relatively pessimistic assumptions, in the future the three richest hydrocarbon states of the Caspian Region would still have total annual revenues from oil and gas of approximately $15 billion, with energy export revenues comprising almost 30 percent of their GDP. As Laurent Rusekas notes, this would place them among the major energy-producing countries, somewhere between Iran and Saudi Arabia.[6]

There is considerable pressure from the oil companies to develop these energy resources relatively quickly. For them, the attraction is especially great since Iran, Iraq, and much of the Persian Gulf have remained closed to exploitation by western multinationals.[7] Led by Chevron, BP Amoco, and Unocal, virtually every major international oil firm has sought contracts in the region: at least 80 oil- and gas-related joint ventures are operating in the region at the present time, with over 30 U.S. companies involved in commercial deals.[8] Rapidly investing $8–9 billion, and promising up to $120 billion in Kazakhstan and Azerbaijan alone by 2015,[9] U.S. companies are seeking a quick competitive advantage. The trend towards mega-mergers in the wake of the 1986 oil price collapse and the Asian crisis is one manifestation of their interest.[10] It makes the companies able to weather the high cost of crude exploitation in this region, and it strengthens them in their bargaining with the countries that are competing very hard for their rigorously prioritized capital budgets. The scramble to sign deals between 1995 and 1998, reminiscent of the behavior of the Seven Sisters a century earlier, is clear evidence that a bonanza is underway.

The pressure to develop resources relatively quickly still exists even though the rush to develop oil and gas reserves has been tempered somewhat by the performance of the international oil market. Because the proclivity of oil prices during the Caspian boom has been to remain relatively low with periodic price spikes, the enthusiasm with which petroleum investments take place tends to ebb and flow in tandem with oil prices. The low prices for crude oil at the end of the 1990s,[11] when combined with the difficulty in locating new finds in the Caspian Basin,[12] dampened the fever to invest which followed the collapse of the Soviet empire. But the almost $30.00 per barrel price of petroleum early in the year 2000 has created a new green light to move ahead quickly. This stop-go pattern greatly complicates the job of Caspian governments. The ability to absorb oil wealth depends in part on whether revenues are generated and introduced in a slow, incremental manner—the optimal scenario that long-time producers before 1974 were able to enjoy—or whether they are the product of relatively unanticipated and especially difficult to manage booms and busts. Like other recent exporters, the countries of the Caspian region fall in this latter category.

The region's geostrategic significance further complicates the picture—and sets the Caspian apart from most other oil exporters. Where the Caspian Basin states differ from oil-exporters like Ecuador is in the importance of their location. This means that pressure to develop oil resources is not merely (or even primarily) economic for some key actors. Because U.S. policymakers view Caspian oil as a potential counterbalance to excessive dependence on the Middle East, especially Saudi Arabia, the United States and its allies claim to have strong national security interests in ensuring that these resources are developed.[13]

But even if oil were not at stake, there are other key interests driving U.S. policy, especially the desire to break regional dependence on Russia, isolate Iran, and support Turkish ambitions in the region.[14] This is a region of strategic importance where Europe and Asia, Christianity and Islam come together—and it has always been the focus of outside interests.[15] Neighboring countries need only be mentioned—China to the east (with its hunger for new energy sources), Russia to the north (with its control over pipelines), and Afghanistan, Iran, and Turkey to the south—to comprehend the extensive interests involved and the desire of big powers to gain influence. Add the irresistible element to this mix of one of the world's largest unexploited sources of oil, and the attention of dominant countries is guaranteed even if other significant alternative energy sources should arise.

More than anything else, the special landlocked status of this region sets the Caspian region apart from all other oil-exporters. Because all other petro-states have access to the sea, no other country has had to enter into such complicated negotiations in order to market its petroleum. Trans-border pipelines are "steel umbilical cords" that link countries together in relatively enduring relationships that last long after the circumstances that led to their construction have changed.[16] Thus, basic issues like preferred routes, timing, and cost have been tremendously politicized. Four major countries—Russia, Turkey, Iran, and China—have sought to build export pipelines across their territory, and at least eight others—the United States, Armenia, Ukraine, Romania, Bulgaria, Albania, Greece, and Macedonia—have sought to exert influence in the controversy over pipeline routes.[17] No other oil-exporter in history has had to maneuver through such a complicated and conflict-laden external environment, and this reinforces a systematic divergence between commercial and political imperatives in the region.

Whatever differences distinguish the Caspian states from other oil-exporters tend to reinforce petrolization, or become less significant in the face of the pressure from the governments themselves to rapidly exploit energy resources. Indeed, this is one of the chief forces fueling petroleum dependence. Facing crises in virtually every arena of governance, rulers need revenues desperately—and very fast—if they wish to hold together their fragile new polities. Like rulers that have experienced energy booms before them, the presidents of Azerbaijan, Kazakhstan, and Turkmenistan (all of whom are former Communist Party heavyweights)[18] must do something to alleviate the deep economic collapse produced during the demise of the Soviet Union if they wish to stay in power. Moribund industrial bases and dying agriculture must be revitalized, new trade links forged in place of the ties to the old Soviet centralized economy, crumbling infrastructure rebuilt, and inflation controlled. The situation of their populations is grim: in Azerbaijan, 61.5 percent of the people live in poverty; 20.4 percent are considered extremely poor; and one in every seven people are refugees.[19] In Kazakhstan, the Red Cross estimates that at least two-thirds of the population live at subsistence levels.[20] And the citizens of Turkmenistan, perhaps the most destitute of the former Soviet subjects, may be even poorer still.

Other pressing priorities push towards rapid petroleum exploitation. Unlike most other oil-exporters with secure borders, Caspian rulers calculate that petrodollars can help them protect the integrity of their newly independent territories from simmering ethnic tensions with and threats from neighboring countries. They must do so (and thereby guarantee their own survival) while trying to diminish the past domination of Russia, threats of radical Islamization, the feared desire of Turkey to extend its borders, and the encroachment of capitalist democracies of the West.

Furthermore, their ecological situation is especially critical, and, in their view, petrodollars might finance some badly-needed clean-up operations.

Pollution threatens potable water sources and fish stocks that provide local jobs; and the rising level of the Caspian Sea poses a dramatic danger to both fixed capital and arable land through widespread coastal flooding.[21] If something is not done, environmental catastrophe threatens to undermine any economic or social progress, with tremendously destabilizing consequences.

In this context, petrodollars are viewed as the badly-needed glue that might hold these polities together. Azerbaijan, which had five presidents between 1991 and 1993 alone, lost 20 percent of its territory in the war with Armenia over Nagorno-Karabakh, and has been flooded with refugees, oil development promises to bring political stability as well as the possibility of regaining land lost in a humiliating defeat. Petrodollars in Kazakhstan could help to paper over internal antagonisms that simmer between ethnic Kazakhs and Russian-Ukrainian Slavs, which have been encouraged by nationalists in Russia who covet the country's northern industrial base. Turkmenistan needs revenues to replace the dramatic collapse of its gas exports when Russia closed its pipeline and to end (or at least diversify from) its reliance on narcotics and arms-trafficking. For Caspian leaders, then, their calculations in the short-term can hardly be anything but energy-based: the faster the exploitation and the more petrodollars available, the better.

Given all of these factors, it is not surprising that both Azerbaijan and Kazakhstan have moved quickly towards dependence on petroleum. In Azerbaijan, oil output is expected to grow from 180,000 b/d in 1996 to 800,000 in 2007. The giant Tengiz project in Kazakhstan has gone from 60,000 barrels per day to about 200,000, and it is projected to pump 700,000 barrels per day at its peak, which should bring in around $5.1 billion annually.[22] Foreign capital is pouring in: in 1996 in Azerbaijan alone, foreign direct investment flows were three times greater than in 1995, the bulk headed for the oil sector. Oil exploitation is slated to provide the state with $100 million per year in the first few years, a figure forecast to increase sharply to more than $2 billion per year from 2006 onwards.[23] In short, regardless of whatever difficulties might plague the oil industry[24] and whatever downward revisions in expectations have taken place, the Caspian Basin has hopes of becoming the new El Dorado.

THE RECORD OF OPEC

As suggested by the real El Dorado, a small poor village at the entrance of the Venezuelan Amazon, the picture may not be as rosy as the current oil euphoria suggests—at least from the perspective of the people who live in the Caspian region. The fate of most of the OPEC countries after the massive booms of the 1970s raises more red flags. The flood of petrodollars in the wake of the sudden 1973 price hike led to the same oil euphoria seen in the

Caspian. Because, by law or custom, oil rents accrued to the state, petrodollars fueled the ambitions of leaders: the Shah of Iran pledged the construction of a "Great Civilization," while across the world President Carlos Andres Perez vowed to build "La Gran Venezuela." As the government revenues of OPEC nations leapt eleven-fold from 1970–1974, petrodollars poured into their national treasuries at an unprecedented rate. This money promised to overcome gaps in national savings, foreign exchange earnings, and national budgets, which are considered some of the chief development constraints. For a period, Caracas and Lagos boasted the same levels of ostentatious wealth that now marks Baku and Ashkhabad.

Twenty-five years after their first boom, however, and despite two other price hikes, most of the OPEC countries are the modern version of the Midas myth. Rather than having met the great expectations raised by oil euphoria, oil-exporters today face a very different reality. Venezuela, Iraq, Kuwait, Libya, and Saudi Arabia have seen real per capita incomes fall to levels not seen since the 1960s, while Algeria, Iran, and Nigeria's incomes have plunged to the levels of twenty-five years ago. OPEC countries are now plagued with double-digit inflation,[25] cost overruns in poorly designed projects, and a monumental waste of resources. Their agriculture and manufacturing sectors foundered, their banking systems collapsed, and their non-oil exports decreased as a percentage of all exports. By the mid-1990s, their annual deficits on goods and services were among the largest of all developing areas; all had become heavily indebted and more dependent on a single export than ever before; and all faced the most serious austerity of the past four decades.[26]

What did OPEC leaders do (or not do) to produce these outcomes? Because all OPEC governments were worried about the exhaustibility of their resources, they declared their intention to "sow the petroleum," that is to create a sustainable basis for a post-oil economy through the promotion of heavy industry, the modernization of infrastructure, and (where necessary) investment in defense. This was accomplished by a conscious policy to expand the jurisdiction of the state as much as possible. While some, like Venezuela, initially seemed to understand that the avalanche of money descending on their states posed dramatic absorption problems and thus set up investment funds to hold petrodollars outside the country, the spending frenzy and rampant rent-seeking that followed the boom ruled out savings. Within a short period these funds were introduced into the economy through government spending; as petrodollar addiction grew, the oil exporters borrowed more and at a faster rate than their non-oil counterparts.[27]

The little we know about how the OPEC states actually spent their money is instructive. Approximately 65–75 percent of the post-1974 gross domestic product was for public and private consumption, largely through subsidies to friends, family, and political supporters of the government. Petrodollars paid for consumer price subsidies on fuel, housing, public services, and utilities. In most countries, subsidies to the private sector were even greater, especially through electricity, transport, communications, tariff barriers, and the like. Former Iranian Finance Minister Jahangir Amuzegar claims that subsidies in the Persian Gulf ran as high as 10 to 20 percent of GDP in some years. The remaining portion (20 to 35 percent of national output) was either invested or used to build sophisticated militaries for national defense or for the suppression of opposition movements.[28] What these figures cannot depict, however, is the untold billions that were wasted or lost through corruption.[29] This spending produced some notable achievements, at least for a time, which were manifest in increased employment opportunities, generous pension plans, and improved public welfare. As a group, the OPEC nations showed a dramatic expansion in infrastructure and allocated a larger share of their national income to education and health than any other group of developing countries.

But soon the cost of overheated economies, widespread inefficiencies, and subsidized living became clear: the average annual real growth of GDP in the twenty years following the 1974 oil boom was actually *lower* than their annual growth rate for the decade prior to the boom. As a result—and in the face of rising birth rates—unemployment and poverty rates soared, price stability and budgetary discipline broke down, borrowing increased, and dependence on capital from oil exports rose.[30] When prices fell in the mid-1980s, these countries were plunged into deep austerity crises from which they still have not emerged. Today, Latin America's two OPEC members, Venezuela and Ecuador, suffer from the worst economic performance on the continent, and even Saudi Arabia, the most well-endowed exporter, suffers from a whopping budget deficit and soaring domestic debt.

Political turmoil accompanied this poor economic record. Regardless of whether prices boomed or went bust, or whether their political regimes were democratic or authoritarian, the abrupt change in circumstances severely tested the polities of all exporting countries. With the exception of Norway, each government had built its political support largely through the lubricant of petrodollar wealth. To the extent that their astuteness in assigning petrodollars had become the basis of their authority, the ratcheting down of oil prices brought instability by exacerbating existing tensions that petrodollars had managed to keep under control. In the earliest and one of the most dramatic cases, the Shah of Iran was overthrown in an Islamic revolution that bitterly criticized the rapid industrialization and westernization of his "Great Civilization." Nigeria oscillated between military and civilian rule, almost in tandem with shifts in oil prices, without being able to consolidate either form of government. As oil prices dropped, Algeria plunged into civil war. Indonesia's Suharto regime, once viewed as an exception to poor management of oil resources, eventually foundered on corruption and crony capitalism. Even Venezuela, Latin America's second oldest democracy, witnessed riots, an attempted military coup, and the demise of its pacted political party system.

Just as gold once tainted everything that Midas touched, oil seemed to drown rather than grease political stability. "It is the devil's excrement," OPEC's founder lamented. "We are drowning in the devil's excrement."[31]

CAN OIL EXPORTERS
(AND THEIR FOREIGN ADVISERS) LEARN?

Can the countries of the Caspian avoid the fate of other oil exporters? Unfortunately, learning seems to be especially difficult where oil revenues are at stake. Few observers seem to grasp the exceptionally close linkages between perverse economic and political outcomes in oil-exporting countries. Instead, economists and experts from international lending organizations consistently attribute poor economic performance to the mismanagement of resources by governments—without realizing that such economically inefficient decision-making is an integral part of the calculation of rulers to retain their political support by distributing petrodollars to their friends, allies, and social support bases.[32] For them, because oil windfalls accrue to the state, the culprit is "big government" and state ownership of resources. Politicians, they argue, planned uncritically, spent their new wealth too rapidly and on the wrong priorities, seemed unable to save abroad, and permitted (or engaged themselves in) excessive rent-seeking activities.

But such observations, while certainly part of the explanation for these perverse outcomes, cannot account for why *all* governments—regardless of the different nature of the regime, different geostrategic considerations, and the very diverse religious, ethnic, and social makeup of their societies—made similar unworkable decisions. Nor can they explain why Norway's state-led development was able to avoid these unfortunate consequences. Finally, they don't take into account the fact that private sectors are just as rent-seeking as political authorities in oil-exporting countries and systematically pressure these authorities to funnel oil money in their direction to finance inefficient and unproductive activities.

Some economists understand that forces deeper than poor decision making are at work and have identified the "Dutch Disease" to explain perverse oil outcomes. Dutch Disease occurs when windfalls push up the real exchange rate, rendering most other exports uncompetitive. As a result, the agriculture and manufacturing sectors of oil countries tend to languish, while the energy sector is usually overdeveloped. Persistent Dutch Disease provokes a rapid, even distorted, growth of services, transportation, and other non-tradables while simultaneously discouraging industrialization and agriculture—a dynamic that policymakers seem incapable of counteracting.[33] The Dutch Disease is especially negative when combined with other barriers to long-term productive activity, such as unstable and often wildly vacillating prices, that are often associated with the exploitation of exhaustible resources.[34] Beginning with Adam Smith, economists have warned of the perils of mineral rents ("the income of men who love to reap where they never sowed").[35] These rents, they argue, too often foster persistent rent-seeking behavior, promote imports rather than food production, foster large-scale but often inefficient models of heavy industrialization, encourage consumption, and generally result in unproductive activities. Vacillating prices make planning difficult and exacerbate unbalanced growth. For them, the roots of misman-agement lie in this "resource curse."[36]

Whether attributed to poor policy decisions or a deeper structural problem, there is remarkable agreement about the prescriptions for avoiding the unfor-tunate fate of most of the OPEC countries.[37] According to economists, the Caspian states should "sterilize" their petroleum revenues by placing them in an oil trust fund abroad, as Venezuela attempted to do, thereby avoiding overly rapid industrialization by introducing them gradually into the economy. These states should use market mechanisms (including a liberalized trade and exchange regime, privatization, and the deregulation of prices, wages, and inter-est rates) to check the role of the state in the economy and to guarantee macro-economic stability as well as a convertible currency. In order to prevent the Dutch Disease, they should improve productivity in agriculture and industry. They must reform the financial sector to increase the independence of the cen-tral bank and strengthen the banking system as a whole. At the same time, they need to set up transparent systems to account for petrodollar spending while simultaneously improving their judicial systems to secure a stable environment of property rights and better fight corruption. In the context of very high poverty levels, they should cut public spending as much as possible and avoid popular public works programs with immediate payoffs. Finally, they should resist the temptation to increase domestic consumption "to placate a restless population" or direct petrodollars to family and friends in favor of concentrat-ing on longer-term goals of raising productivity. If the governments of Central Asia do not follow these prescriptions, they are warned, the past is an eloquent mirror of their future. OPEC is an omen, Amuzegar writes, and the Caspian states "would do well to learn from their predecessors' failures."[38]

But learning these lessons—and applying these sorts of policies—is espe-cially difficult in petro-states. Basically, these states are being asked to remake themselves—and quickly—while virtually all incentives push their leaders in a different direction. Not only are billions of dollars up for grabs, but they generally circulate in the context of weak administrative structures, insecure property rights, nonexistent judicial constraints, deep divisions, and strong political ambitions. Successful efforts to use petrodollars wisely depend, above all else, on the presence of countervailing political and social pressures strong enough to curb what I have elsewhere called "petrolization," a process by which states become dependent on oil exports and their polities develop an addiction to petrodollars.[39] Such countervailing pressures include capable state administrative structures with developed civil services and non–oil-based tax systems, economic activities and social forces that are independent from mineral rents, and, perhaps most important, transparent democratic institutions which are powerful enough to rein in the alliance between multinational oil interests and political leaders.[40] These factors can constrain the centralizing and concentrating tendencies that petroleum exploitation exerts on the polity and limit the powerful alliance between rulers and oil companies that initially takes place when oil revenues circulate without strict controls.[41]

Where these institutions do not exist, the learning that takes place may not be the sort that social scientists recommend or expect from "rational" actors. Rather than avoiding the hasty industrialization, profligate overspending, and increased domestic consumption that marked most OPEC countries (as development economists advocate); or checking the rising dominance of the state over the economy (as neo-liberals advise); or promoting judicial reform, financial transparency and "good governance" (as both USAID and the World Bank urge), political leaders may find that the immediate political and economic incentives they face encourage them to do precisely the opposite. Indeed, it is precisely the structure of choice in oil-exporting countries—both constructed by and subsequently based upon the highly politicized allocation of rents—that renders many of the prescriptions of economists unlikely. This is not because leaders do not understand what might be in their own interests; rather, at least in the short run, they understand all too well!

Without strong countervailing pressures, the incentive structure for both politicians and the private sector in oil-exporting countries is perverse. Petro-states are not like other states. While they share many of the development patterns of other developing countries, especially mineral exporters, the economies and polities of countries dependent on oil are rapidly and relentlessly reshaped by the influx of petrodollars in a manner that sets them apart. This is especially true if petroleum exploitation coincides with modern state-building, as it does in the Caspian region. Where this historical coincidence occurs, petro-states are marked by especially skewed capacities. In effect, their frameworks for decision making are altered in a manner that encourages mineral-based development, numerous institutional inefficiencies (especially the lack of a tax system), the concentration of power, and, therefore, the substitution of money for statecraft.

What distinguishes oil states from all other states, above all else, is their addiction to oil rents. Where this oil addiction takes hold, a skewed set of both political and market incentives so penetrate all aspects of life that almost anything is eventually up for sale. Actors in oil states do not behave the same as they do elsewhere; they simply do not have to. Thus oil companies do not assess political risk in the ways that many other firms do. (Is it even conceivable to picture manufacturing exporters competing for entry into the Caspian

Sea region as the oil companies are doing, seemingly undaunted by surrounding conflicts in Chechnya, Afghanistan, Iraq, and Armenia? Or is it possible to imagine many other industries withstanding the systematic bombing of their infrastructure as the oil companies do in Colombia?) Nor do domestic entrepreneurs, labor leaders, or, most especially, political leaders make their calculations in the same way. Where oil is the focus of both domestic and international competition, the stakes are simply too high.

In this respect, petro-states are rentier states *par excellence:* not only does petroleum provide exceptionally high levels of rents over a long period of time, but it also facilitates international borrowing, thereby perpetuating this capacity to live beyond their means. This permits the leaders of petro-states to avoid badly needed structural changes far longer than those in other developing countries, which are reined in more quickly when their macroeconomic indicators show trouble. But with oil as collateral, petro-states are seldom forced to adapt—and never on time. Nor are they required to tax their own citizens, which is perhaps the key mechanism eventually producing the accountability of governments. This is a short-term boon to policymakers, but problems that are constantly deferred pile up and worsen. Thus change, when it finally does come, is likely to be dramatic and not incremental, driven by profound political and economic crises.

Furthermore, all petro-states have a similar self-reinforcing dynamic. Both private economic power and political authority rest on the dual capacity to extract rents externally from the global energy environment instead of from their population and to distribute them internally using political criteria as the central mechanism of allocation. This creates an exceptionally close linkage between economic and political power, and develops networks of complicity based on the classic exchange between the right to rule and the right to make money. It also tightly links economic and political outcomes in a manner akin to former socialist countries—a reality that seems to elude most observers. Instead, economists and experts from international lending organizations consistently attribute the poor economic performance of rich oil exporters to the mismanagement of resources by governments or to the overextended role of the state, and they often see little relationship between economic performance and dramatic political changes inside countries like Iran, Venezuela, Indonesia, Algeria, or Nigeria.[42] They do not realize that such economically inefficient decision-making is an integral part of the calculation of rulers to retain their political support. Nor do they recognize that economic reforms can seldom be accomplished short of a regime change—and even this does not guarantee the sort of policies that social scientists recommend or expect from "rational" actors.

In effect, rulers of oil-exporting countries have no immediate incentives to be frugal, efficient, and cautious in their policy-making, and they have no reason to decentralize power to other stakeholders. To the contrary, revenues pouring into a highly concentrated structure of power lead to further concentration and encourage further inefficient and unproductive spending to establish and maintain rentier networks between politicians and capitalists.

OPEC LESSONS FOR THE CASPIAN

Will the countries of the Caspian region follow the OPEC example? Should we expect to see the economic decline, political decay, and (in some cases) heightened regional turmoil that now mark even some of the capital surplus oil-exporters of the Persian Gulf? The International Monetary Fund, some government agencies, and some private observers have emphasized that the progress made in the Caspian's transitional economies and the lessons learned from other exporters belie this scenario; they stress the commitment of Caspian governments to a diversified market economy.[43] While some of this analysis may be motivated by other than impartial and disinterested criteria,[44] there is no question that Caspian countries have had some impressive successes. Perhaps most important, they have established macroeconomic stability after some rocky years. Azerbaijan, for example, went from 1800 percent inflation in 1994, the year it embarked on a stabilization program, to 5 percent in 1997. Meanwhile, Kazakhstan's annual inflation rate fell to less than 10 percent from a high of 2000 percent in 1994.[45]

Under the influence of the International Monetary Fund (IMF) and other agencies that embrace neoliberal economic policies, Caspian leaders have also sought to avoid, primarily through privatizations, the sharp expansion of the state that characterized other oil-exporters in the midst of booms. This has proceeded farthest in Kazakhstan, which unlike its neighbors moved significant oil and power generation into private ownership beginning in 1996 (until 1998 when privatization was suspended). It has paid off in the short run by inviting an unprecedented level of direct foreign investment in all areas of its energy sector, making Kazakhstan the largest recipient of foreign direct investment of all the members of the Commonwealth of Independent States (CIS).[46] The government also privatized small and medium-sized enterprises, with preference given to domestic investors.[47] In Azerbaijan, the private sector's share of GDP has risen from 25 percent to 40 percent in the past year alone. While Turkmenistan has not made similar headway, the government has approved a socioeconomic strategy that includes "measured" steps towards marketization.

Progress does not stop there. To counteract the windfall problems characteristic of other oil-exporters, the IMF has insisted that Azerbaijan establish a special account at the central bank for early-oil bonus payments; according to this plan, money will be "sterilized" in accounts abroad and only gradually fed into the budget on a predetermined schedule, thus enabling the country to avoid the overspending and ratchet effects that proved so devastating to OPEC.[48]

Other efforts have been made to counteract the ill effects of petroleum-led development. For example, Azerbaijan has committed itself to slowing real exchange rate appreciations by eliminating formal trade barriers and implementing tight fiscal policies. In addition, banking reform is promised in Azerbaijan, which is still dominated by the four large state-owned banks that are a legacy of the Soviet system.[49] In order to accurately trace financial flows, financial sector reform has proceeded apace in Kazakhstan, where institutions are the strongest in the region. Property rights, at least with regard to petroleum, seem more secure from the companies' point of view, and the investment climate has improved. The terms of contracts, often drafted by oil company lawyers, rule out fundamental renegotiations and nationalization.[50] Finally, both Azerbaijan and Kazakhstan have implemented important new tax codes, which were initially considered to be a big step in the right direction since they provided for uniform taxation with only a limited set of exemptions.[51]

Nonetheless, signs are not promising that the Caspian region can avoid the painful outcomes of most OPEC countries. First, the past experience of oil exploitation in the region, even prior to Soviet domination, does not bode well. The presence of oil and gas in the Caucasus and Central Asia is nothing new: indeed, there are recorded reports of petroleum as far back as the thirteenth century, and at the turn of the last century, the wells of what is now Azerbaijan pumped out about half of the world's oil supply. Yet there is little indication that oil wealth ever helped to build viable polities and economies. Instead, the fruits of past booms were gathered by the great oil barons of the day—the Nobel brothers, the Rothschilds, and the leaders of Royal Dutch/Shell—and largely did not accrue to the majority.[52]

Russian domination perpetuated poor economic outcomes, though the causes and mechanisms were different. As an administrative sub-unit of the Soviet Union, oil and gas production remained the linchpins of the Azeri economy under a hyper-centralized system that allocated these tasks to the Caspian. While the patterns were different from the neo-colonial status of other oil-exporters, which exported raw materials in exchange for the consumer goods and industrial equipment they needed, economic development was still highly skewed. In the Azerbaijani Soviet Socialist Republic (AzSSR), Baku's two refineries processed close to 400,000 barrels of crude oil per day, about half of it imported from Russia, Kazakhstan, and Turkmenistan; and industrialization revolved around oil production equipment for use in West Siberia.[53] But by the 1970s, Azerbaijan was stagnating along with the other Caspian states. According to John Willerton, energy development produced the lowest rate of industrial growth of any Soviet Republic since the end of World War II.[54] Thus, even though Soviet-directed economies were foundering virtually everywhere, energy-led development added a special dimension to this demise and helps to explain why economic disintegration was even deeper here.

The region's historical experience also sheds light on the debate over whether petrodollars might alleviate tensions, which John Roberts calls "perhaps the basic premise of Western activities in developing the energy resources of the Caspian"[55]—or whether they might produce the opposite result. Especially during both world wars, the mineral wealth of the Caucasus did little to bring independence and stability; to the contrary, it became the key source of contention between external superpowers, drawing the region into their conflicts.[56] More important, the fierce competition between the companies and the corruption that ultimately swept the territory fueled existing ethnic divisions to dangerous levels. The strategic oil calculations of others exacerbated tensions between the Turkic and Armenian ethnic groups, and created some of the roots of the conflict that has erupted today between Azerbaijan and Armenia over Nagorno-Karabakh. Stalin's fateful decision to overturn that area's possible unification with Armenia and transfer jurisdiction to Azerbaijan, for example, was partially motivated by control over mineral wealth. Such calculations continue today, especially with regard to export pipeline politics in this landlocked area, and they influence the prospects for sustaining a fragile ceasefire between Azerbaijan and Armenia. Because the continuation of the war renders some non-Russian pipeline routes useless, this gives Russia an advantage in securing its own goals for Caspian oil to be transported only through Russian domestic pipelines. Commenting on how brutal struggles over ethnicity and sovereignty have consistently been intertwined with petroleum, Rosemary Forsythe warns, "The dangers of historical parallels notwithstanding, there are clear similarities between then and now—particularly in commercial competition, corruption, poor administration and ethnic tensions."[57]

But even if this history is ignored, the current signs are not good. As oil euphoria has shot up, signs of the Dutch Disease are already evident. In Azerbaijan, oil exports, which made up only 33.9 percent of total exports in 1994, soared to a stunning 66.4 percent a mere two years later.[58] This new oil dependence has already resulted in a structural change in the economy that favors services and trade and is biased against industry and agriculture. The production of virtually all non–oil-related industrial items has slipped substantially between 1992–1996;[59] exports of non-oil-related products, like textiles and food products, have dropped respectively from 18 percent and 10 percent of total exports in 1994 to 10.8 percent and 4.4 percent in 1996; and imports of food products have soared from 26.3 percent of total consumption in 1994 to an unprecedented 60.7 percent in 1996.[60] In Kazakhstan, the rise in energy production in 1997 from increased outputs at the Karachaganak field, and later Tengiz, and the prospects of plenty more revenues in the future, have encouraged overspending for consumption and shifted attention away from the potential development of other sectors, especially agriculture.[61] The same is true for Turkmenistan, which has been scrambling for a pipeline (going in any direction) as its access to Russian export pipelines declines. While these developments mirror in many ways the decline in agriculture and the shift towards food imports that has plagued some non-oil post Soviet states, oil has exacerbated this process.

Tax developments are even more worrisome. Because tax laws are one of the central mechanisms for either reinforcing or combating petrolization, and because they are so crucial for building capable and efficacious states,[62] the trends here are especially revealing. In Kazakhstan, Richard Auty notes that the mere expectation that the country would someday be a major energy producer and be able to rely upon oil and gas revenues means that officials have been reluctant to develop a reliable system of tax collection. Instead, even when state revenues began to plunge and tax revenues as a percentage of GDP declined through 1998, officials relied upon the hope of future petrodollars and borrowed rather than improve tax collection—even though some estimates of the oil rents Kazakhstan is likely to receive may be less than what the government could achieve merely by improving its tax collection efficiency.[63]

In Azerbaijan, the oil bias in the taxation process is also evident, although the dynamics are different. Changes in the tax system indicate a willingness to sacrifice investments in non-oil sectors of the economy in order to attract more investment into the offshore hydrocarbon operations. In effect, there are two tax policies at work: the statutory regime and the offshore oil consortia regime.[64] The latter permits oil companies to be shielded from statutory taxation while other investors are subject to the old Soviet laws still on the books.[65] Azerbaijan also imposes a relatively small tax burden on oil producers compared to oil-exporters outside the former Soviet Union.[66]

Decision making biased towards the development of the oil industry and the attraction of necessary foreign investment in that sector represents a clear choice made by rulers. But given the tremendous crises facing these new states and the opportunities presented by their energy resources, what else could they have done? What other development strategy could possibly have looked as attractive in the short run? The calculations of rulers are not based on possible long-term payoffs—to be enjoyed by some other president long after they are gone. For the leaders of the Caspian states, petrolization, regardless of its costs, has important political advantages—at least in the short to medium term. Petrodollars can fortify their own rule, expand the reach of the executive, and concentrate power at the top. At least initially, development based on energy resources produces a classic alliance between foreign companies and local rulers to sustain each other's interests. Thus the three presidents of Azerbaijan, Kazakhstan, and Turkmenistan are the principal partners of more than 50 major energy companies and other multinationals, including Unocal, Bechtel, Chevron, Elf, Total, British Petroleum, and Royal Dutch/Shell. Looking out for themselves means looking out for the companies.

The exploitation of energy resources initially supports whatever regime type is already in place when new revenues come on stream.[67] In the Caspian, these regimes are autocracies, built upon long-standing practices that pre-dated Soviet domination and were reinforced by it. In Azerbaijan, power is strongly concentrated in the executive. In part due to the fact that the war in Nagorno-Karabakh coincided with Azeri independence, both presidential and military power developed with greater speed than the rest of the government's functions; and the Parliament became a rubber stamp body that has lit-tle influence over policy-making.[68] Kazakhstan's Nazarbaev also exercises absolute authority—rigging elections, dissolving the independent-minded constitutional court, barring oppositions from running for office, dictating policy in all areas of government, maintaining an extensive surveillance appa-ratus, and installing a new constitution which firmly entrenches the power of the executive. In Turkmenistan, President Niyazov, who instituted a person-ality cult and renamed himself "Turkmenbashi ("Chieftain of the Turkmen"), controls the only Turkmen language television station and has extended his rule until 2002 without an election.[69]

Furthermore, oil revenues have fortified some of the most unfortunate legacies of Soviet rule. Although Central Asia and the South Caucasus were sub-units of a larger system, power was always centralized and concentrated towards the top: but today this is even more true. Communist bureaucratic patterns are still in place; in effect the basic elements of the command-administrative system in each republic have been preserved virtually without alteration. These bureaucratic structures are tightly linked to the individuals that head them, and the leaders of these bureaucracies, in turn, are appointed by the President. Oil revenues, which are allocated primarily by the president, strengthen this command structure by greasing the wheels from the top. The success of economic reforms notwithstanding, the tendency towards the expansion of the public sector and increased state intervention into all aspects of life has not been eradicated; it has only changed its form. Finally, corrup-tion, which was a pervasive part of the Soviet system and which has deep roots in the region—especially since illegal economic activity seems to have been most prevalent in the southern republics—appears to have increased dramat-ically. According to Rasizade, this situation has produced a mentality of "per-sonal profiteering by those with political power" as well as a tendency to see a country's wealth as "a pool from which to draw individual fortunes."[70] The state, always the source of illegal enrichment in the past, has become the equivalent of a huge milk cow.

Other examples abound demonstrating how oil dependence reinforces authoritarian legacies, and is reinforced by them. In each country the power of the executive has increased since independence because the president has acquired the right to set the rules of the game for international oil contracts, and they have sought the active support of foreign capital in the concentration of their own authority. President Aliyev of Azerbaijan has designed tax laws that give the top echelons of government maximum control over the awarding and subsequent distribution of oil rents. Thus the ratification of each new offshore contract allows the government to negotiate a new tax regime, increasing its own leverage.[71]

President Aliyev's son is first vice president of the State Oil Company of the Azerbaijani Republic (SOCAR), which has the sole authority to register all newly discovered commercially feasible sites, and SOCAR officials are appointed by presidential decree. The president authorizes all contracts in order to make sure the right companies from the right countries are given the right share in emerging oil consortiums, thus the companies must deal directly with him. He has persistently extended protection to oil executives from the West in exchange for favors from them. This alliance has fortified a pre-existing network of nepotism and corruption extending to all areas of Azeri life, producing a "hijacking of assets and wealth by a few people while a large part of the country goes empty-handed."[72] Not surprisingly, the government has decided that oil will not be privatized; because such strong rent-seeking incentives exist, another ruler is unlikely to undo this decision.[73]

The pattern is similar in Kazakhstan. As a means of backing away from the consequences of privatization of energy resources, since 1997 the government has made all oil contracts subject to statutory evaluation by the State Tax Committee (which is conveniently headed by a relative of the president). Each contract must be made directly with the government or some other state body, and the law requires that the government determine the type and level of taxes and surcharges applicable in each case. Not surprisingly, there are no provisions to ensure an open and fair competition for these incentives, and the State Tax Committee has been criticized for their lack of transparency. Because the lack of clarity in tax laws encourages "creative" interpretations from the tax police and others of the actual taxes owed by each company, corruption abounds. In one case, a $500 million payment made by Mobil for some of its share in the Tengiz oil field allegedly never reached the budget. With President Nazarbaev's daughter running the state television company, his son-in-law heading the tax police, and a relative directing the powerful foreign investment committee, complaints about these practices are virtually non-existent.[74]

Buttressing this classic tit-for-tat relationship between top leaders and oil companies is the lack of any democratic accountability. Elections in these three Caspian states are neither free nor fair, and opposition groups are routinely excluded from participation. In 1998, for example, when Azerbaijan's Aliyev stood for reelection, he won with a suspicious 75 percent of the vote; there were widespread reports of fraud, and one electoral district announced a voter turnout well in excess of 100 percent![75] Governments control the media, and there is a lack of basic freedoms in virtually all areas. Freedom House, which rated Azerbaijan as "not free," notes that the country has earned one of the lowest freedom scores in the world—with the rest of the Caspian region not far behind.

Institutional arrangements of this sort indicate that mechanisms of accountability, which seem to have the prospects for reining in petrolization, are unlikely to work—no matter how good they look on paper. Take the special oil account of Azerbaijan, for example, which is thus far the only effort aimed at sterilizing petrodollars and the single most important reform for preventing the excesses of petroleum-led development. When viewed from this political perspective, it has a fatal flaw. Because the oil account involves profit-sharing between the central bank and the government at the same time that it relies on the autonomy of the bank to restrain the government, it not only blurs the functions of the central bank but it also pits the bank against the executive branch in a context in which the bank has little independence. Thus, while the IMF seems to have pushed authorities to establish a transparent institutional mechanism that could protect oil wealth for the future and has also argued for an oil trust fund managed outside the budget to be held outside Azerbaijan, there is no real indication that the government would favor such a significant loss of control (despite rhetoric to the contrary) or examine the anti-corruption measures that might serve as constraints against rent-seeking in Azerbaijan and Kazakhstan. In the few cases where arrests are actually made, they generally involve middle-level criminal groups of state officials, police officers, and ordinary gangsters; suspected large-scale embezzlers, like the deputy chief of Azerbaijan's fuel distribution concern (Azeravtoyal), always manage to escape punishment.[76]

In this institutional context, the countries of the Caspian are likely to resemble their OPEC counterparts in their overspending, overborrowing, general inability to manage oil rents, and rampant corruption. Their basic pattern of rent-seeking should also look similar: they too will maximize their capacity to extract from the oil companies by playing different interests against each other while distributing these rents as the key mechanism for remaining in power. In the context of the unpredictability of oil revenue inflows, this should mean papering over the current account deficits with debt when times are tough, then trying to pay back these debts during booms when revenues are desperately needed elsewhere. Such actions will be necessary, not only to assuage creditors and rent-seekers and keep political tensions within the regime at bay, but also to show tangible improvement to the number of people living at subsistence levels and prepare for (or wage) war against neighbors or ethnic groups. Given these enormous pressures, prospects for holding back spending and avoiding the "resource curse" are poor.

CONCLUSION

What steps could be taken to change the relentless political and economic incentives encouraged by oil booms? Externally, a united effort on the part of governments and lending agencies to make future borrowing and aid contingent upon the establishment of capable and transparent tax and administrative structures is essential. Internally, weak opposition groups and nongovernmental organizations should be fostered, especially since accountable and transparent institutions tend to be produced when power is relatively dispersed. Finally, the active presence of human rights and environmental organizations needs to be encouraged, particularly the latter which can draw upon support from fishing interests in the Caspian Sea. Together these diverse forces might be able to construct a transnational alliance for curbing the excesses of oil-led development.

But even if all of this should occur, the incentive structures created by centuries of despotism and domination and then remolded by sudden oil wealth will not easily be changed. The prospects of significant oil and gas wealth have raised very high expectations in the Caspian states, partly due to the rhetoric of the governments of the region. If prosperity does not materialize over time, and populations believe that they have endured booms and busts only to witness the enrichment of a small elite, the consequences could be extremely serious—as illustrated by the turmoil in oil-exporters from Venezuela and Ecuador to Algeria, Indonesia, and Nigeria; indeed, some observers are already predicting an anti-western backlash or an Iranian-type Islamic revolution in the region, as well as increased interstate warfare.[77] But whatever the ultimate outcome, one result is easy to predict. If oil-fueled rentier behavior becomes the entrenched norm in both the public and the private sectors over time, market efficiency, state capacity, and democratic accountability are likely to fall by the wayside in the face of the crude calculations of elites.

NOTES

I am grateful for the invaluable comments and assistance of Inna Sayfer and Laurent Ruseckas, who taught me about the countries of the Caspian Basin, but who bear no responsibility for my errors of interpretation. I also owe special thanks to the European University Institute, especially Philippe Schmitter and Eva Breivik.

1. See Frank Viviano, "Perilous Lifelines to West: Conflict-Ridden Caspian Basin is the World's Next Persian Gulf," *San Francisco Chronicle,* August 10, 1998, 1. For other discussions in this vein, see Frederick Starr, "Power Failure: American Policy in the Caspian," *National Interest* 47, Spring 1997: 20–21; and Chris Kutschera, "Azerbaijan: The Kuwait of the Caucasus?" *Middle East* (London), March 1996.

2. The categories of *capital deficient* and *capital surplus* oil exporters are generated by examining the relationship between the populations of oil-exporting countries and their projected oil reserves. Because estimates of the amount of petroleum in the Caspian Basin vary substantially, as we shall see below, their categorization is not yet certain. For more on this point, see Terry Lynn Karl, *The Paradox of Plenty: Oil Booms and Petro-States* (Berkeley: University of California Press, 1997), 17.

3. This remark by Valekh Aleskerov is cited in Andrew Meier and Scott McCloud, "Caspian Black Gold," *Time International* 150, no. 44 (June 29, 1998).

4. Ian Bremmer, "The Riches of the Caspian," *World Policy Journal* 10, no. 1 (Spring 1998). Whether conservatively appraised at 30 billion barrels (80 percent of which may lie in Kazakhstan), the 50 billion barrels estimated by petroleum industry analysts, or measured at the overly optimistic number of 200 billion barrels used by the U.S. Department of Energy and the Caspian governments themselves, Caspian Sea reserves are significant. The British Petroleum Statistical Review cites Azerbaijan's reserves at only 7 billion and those of Kazakhstan at 8 billion, but U.S. sources have consistently talked of much bigger reserves. Rice University's James Baker III Institute for Public Policy puts reserves at somewhere between 15–30 billion barrels of crude. This latter figure is about 3 percent of the world's oil supply—compared to about 60 percent of known reserves in the Middle East. *Wall Street Journal*, October 12, 1998.

5. If predictions prove correct regarding a rise by at least one-third in the global demand for petroleum by 2010, with a corresponding rise in total natural gas consumption, the supply of conventional oil is likely to be unable to keep up with demand. According to the International Energy Agency, by 2010 total demand will be between 92 and 97 million b/d. At present, world demand is rising at about 2 percent per year. But this gradual rise could be offset by efforts at conservation, new discoveries, technological innovation, or protracted economic stagnation, especially in Asia. See "Central Asia Survey," *The Economist*, February 7, 1998, 5; and *Wall Street Journal*, January 18, 1999. A good explication of a possible trend of rising prices can be found in Colin J. Campbell and Jean H. Laherrere "The End of Cheap Oil," *Scientific American*, Special Report: March 1998, 78–83; or Daniel Yergin and Dennis Eklof, "Fueling Asia's Recovery," *Foreign Affairs* (March–April 1998). For the counter argument, see Amy Myers Jaffe and Robert A. Manning, "The Shocks of a World of Cheap Oil," *Foreign Affairs* 79, no. 1 (January–February 2000).

6. Laurent Ruseckas, "State of the Field Report: Energy and Politics in Central Asia and the Caucasus," *Access Asia Review* no. 2 (The National Bureau of Asian Research): 77.

7. Even though Iran has signed deals to develop an offshore gas field and is currently in discussions with a number of companies, U.S. based companies are barred from these transactions and threats of U.S. sanctions have delayed the negotiations of other companies. Iraq is still closed for business until sanctions are lifted.

8. The five largest development projects are: the Kazakhstan-Chevron joint venture "Tengizchevroil," the Azerbaijan International Operating Company (AIOC), the Azerbaijani Karabakh field in the Caspian, Kazakhstan's Karachaganak oil and gas field, and some projects in Kazakhstan's offshore area. See Rosemary Forsythe, "The Politics of Oil in the Caucasus and Central Asia," *Adelphi Papers*, no. 300 (1996), chap. 3; and Viviano, "Perilous Lifelines to the West."

9. This figure applies only if every undrilled structure offshore turns out to be a large oil field (which they won't). Ruseckas estimates that real investments by 2005 are likely to be an additional $15 billion. Laurent Ruseckas, correspondence with author, April 26, 1999.

10. As Daniel Yergin points out, greater scale enables companies to carry projects like the Caspian Sea on their portfolio and carry them out over a number of years despite the pressures caused by the volatility of prices. It also enables them to combine capabilities in oil and natural gas, an ability which will be important in this region. See "Can Big Oil Survive Cheap Oil?" *WSJ Interactive Edition of the Web*, December 10, 1998.

11. Production costs for Caspian Sea oil should be around $8 per barrel, as compared to a mere $3 per barrel in Saudi Arabia. Thus Caspian reserves could be uneconomic to produce if oil prices fall below $9 per barrel. See Michael Page, "A Little Less Sizzle in the Region's Expectations," *Petroleum Economist,* September 1998.

12. No significant new oil reserves have been found offshore Azerbaijan since the collapse of the Soviet Union. Discouraged, some companies are leaving, and the Caspian International Petroleum Company (CIPCO), a consortium led by Pennzoil, has even shut down because it failed to find enough oil to justify full-scale drilling. Nonetheless, both Norway's Statoil and Exxon are continuing to increase their investments in the region. See *The Economist,* March 13, 1999, 86; and *Wall Street Journal,* October 12, 1998.

13. As U.S. Secretary of Energy Bill Richardson has noted, "The Caspian region will hopefully save us from total dependence on Middle East oil." See Jeffrey Goldberg, "The Crude Face of Global Capitalism," *New York Times,* October 14, 1998. Numerous scholars and policy makers have also noted that Central Asian oil has the potential to offer an important alternative to the Middle East, thereby permitting a diversification of supply. This is especially important because even greater U.S. dependence on the Persian Gulf in the next century has been widely predicted. See especially Daniel Yergin and Joseph Stanislaw, "Oil: Reopening the Door," *Foreign Affairs* 72, no. 4 (September 1993).

14. Ian Bremmer notes that opposition to the expansion of Iranian interests; a desire to promote the interests of Turkey, a NATO ally and secular model for Islamic nations; and a somewhat contradictory policy towards Russia have determined much of U.S. government actions in the region. As a committee set up by President Clinton to oversee the interests of U.S. oil companies in the Caspian has made clear, U.S. policy will be linked to these issues as well as to tension between Pakistan and India, China's future policy towards its neighbors, and the potential spread of Islam in the region. Because the Caspian reserves offer the United States the enticing prospect of political leverage in the region in the post-Soviet era, political consideration will determine Washington's agenda as much as the search for guaranteed hydrocarbon deposits. Bremmer, "The Riches of the Caspian."

15. The region known as the Transcaucasus (the isthmus south of the Caucasus Mountains and between the Black and Caspian Seas) is a strategic land bridge between Asia and Europe, and it has persistently been invaded over the centuries by Romans, Arabs, and Turks. The Russian Empire sought to control the region in the early 1700s, which was subsequently disputed in a series of wars between Russia, Turkey, and Iran. Michael P. Croissant, *The Armenia-Azerbaijan Conflict: Causes and Implications* (London: Praeger, 1998).

16. This phrase is quoted in Ruseckas, "State of the Field Report," 51.

17. Forsythe, "The Politics of Oil in the Caucasus and Central Asia."

18. Azerbaijan's President Heydar Aliyev is a former Communist Party secretary and KGB chief in Baku; Kazakhstan's president Nursultan Nazarbaev is a former member of the Soviet Politburo, and Turkmenistan's president Saparmurat Niyazov is the former chairman of the Supreme Soviet in Ashkhabad.

19. The Azerbaijani government estimates the figure to be nearer to 90 percent. See Economic Intelligence Unit, *Country Profile: Azerbaijan, 1998–1999,* 17.

20. According to the IMF, in Kazakhstan, during the initial stages of its transition from Soviet rule, real GDP declined by 31 percent from 1992 to 1995. International Monetary Fund, "IMF Concludes Article IV Consultation with Kazakhstan," Press Information Notice, June 24, 1998.

21. Douglas Blum, "The Russian Trade-off: Environment and Development in the Caspian Sea," *Journal of Environment and Development* 7, no. 3 (September 1999); and Robert Cullen, "The Rise and Fall of the Caspian Sea," *National Geographic* 195, no. 5 (May 1999). For more on the priorities of these governments, see the chapters by David Hoffman, Nancy Lubin, and Pauline Jones Luong in this volume.

22. Chevron is planning to invest over $20 billion into Tengiz over the next 40 years in expectation of far greater profits. See "Big Oil's Pipe Dream," *Fortune*, March 2, 1998.

23. Michael Wyzan, "Oil Price Decline Fails to Dampen Azerbaijan's Recovery," *Radio Free Europe/Radio Liberty (RFE/RL)*, June 29, 1998.

24. One of the main difficulties is the lack of export pipelines. For a brief but excellent discussion of the various options for pipelines, see "Caspian Gamble: A Survey of Central Asia," *The Economist*, February 7, 1998, 7–10; and Paul Sampson, "Lubricating the Caspian," *Transitions*, February 1999.

25. The small Persian Gulf monarchies are an exception here.

26. Karl, *Paradox of Plenty*, 1997, 25–32; and A. Gelb, *Oil Windfalls: Blessing or Curse?* (New York: Oxford University Press, 1988).

27. Karl, *Paradox of Plenty*, 1997, 29.

28. Jahangir Amuzegar, "OPEC as Omen," *Foreign Affairs* 77, no. 6 (November–December 1998): 101.

29. One manifestation of the chaos caused by oil booms is that there is no accounting of the utilization of oil windfalls by the OPEC countries themselves. The figures used here are estimates by Gelb, *Oil Windfalls;* and Amuzegar, "OPEC as Omen."

30. For more on these outcomes, see Karl, *Paradox of Plenty*, 1997, chapter 2.

31. This is Juan Pablo Perez Alfonzo's description of the impact of petroleum. He is widely credited with being the founder of the Organization of Petroleum Exporting Countries (OPEC). Interview with author, Caracas, summer, 1976.

32. Jahangir Amuzegar, for example, writes: "Apart from a number of traumas unrelated to oil—a revolution in Iran, two bloody and ruinous wars between Iraq and its neighbors, and coups in Nigeria, Qatar and Venezuela—the OPEC members' own miscalculations and mismanagement ultimately brought them external payment deficits, rising budgetary shortfalls, runaway inflation . . . " Amuzegar, "OPEC as Omen," 99.

33. W. Max Corden, "Booming Sector and Dutch Disease Economics: A Survey," *Working Paper 079* (Canberra: Australian National University, 1982); and Peter J. Neary, ed., *Natural Resources and the Macroeconomy* (Cambridge, Mass.: MIT Press, 1986).

34. T. J. C. Robinson, *Economic Theories of Exhaustible Resources* (London: Routledge, 1989).

35. Adam Smith, *An Inquiry into the Nature and Causes of the Wealth of Nations* (New York: Modern Library, 1937).

36. For an extensive discussion of the "resource curse," see Richard Auty, *Sustaining Development in Mineral Economies: The Resource Curse Thesis* (London: Routledge, 1993); and Robinson, *Economic Theories of Exhaustible Resources*. For its application to oil exporters, see Karl, *Paradox of Plenty*, chapter 3. Auty's more recent work demonstrates how resource endowments make it difficult to transform industrial policies because they build a type of inflexibility into the economic structure.

Less-well-endowed countries, to the contrary, are able to change their industrial policies more easily, thus laying the basis for greater competitiveness, improved foreign exchange earnings, and greater economic growth. See Richard Auty, "Industrial Policy Reform in Six Large Newly Industrializing Countries: The Resource Curse Thesis," *World Development* 22, no. 1 (1994): 11–26.

37. See, for example, the overlap in prescriptions between Amuzegar in "OPEC as Omen," and Christoph Rosenberg and Tapio Saavalainen in "Dealing with Azerbaijan's Oil Boom," *IMF Working Papers* 35, no. 3 (September 1998).

38. Amuzegar, "OPEC as Omen," 95.

39. Karl, *Paradox of Plenty*, 1997, 44–70.

40. This argument is developed in Karl (1997) more completely than it can be presented here. It is based on the experience of the capital-deficient oil-exporters, however, and does not include the countries of the former Soviet Union.

41. It is important to note that eventually this alliance develops serious strains, generally after oil-led development has produced a burgeoning middle class, but this may take decades. The best description of the changing alliance between companies and rulers still remains Franklin Tugwell, *The Politics of Oil in Venezuela* (Stanford, Calif.: Stanford University Press, 1975), which describes how, over time and with the development of new social sectors, agreements between political leaders and multinational oil companies eventually develop into more complicated negotiations over petroleum rents.

42. International oil economist Amuzegar ("OPEC as Omen," 99), for example, does not draw this connection. See note 32.

43. For example, IMF materials claim that the government of Azerbaijan recognizes the need to ensure that non-oil sectors of the economy grow, and they point to plans to further liberalize exports, free prices, and continue a virtually halted privatization program. International Monetary Fund, "Azerbaijan Enhanced Structural Adjustment Facility and Extended Fund Facility Policy Framework Paper, 1997–2000," online at <http://www.imf.org>.

44. For well-connected Republicans, for example, the Azerbaijan lobby has been a gold mine. Peter Stone reports that former Secretary of State Lawrence Eagleburger and former Senate Majority Leader Howard Baker represent Azerbaijan for $40,000 a month. Stone also paints an impressive picture of the business interests at work to polish the government's image as well as to overcome the strong Armenia lobby in Washington. See *The National Interest* 31, no. 11 (March 13, 1999).

45. "IMF Concludes Article IV Consultation with Kazakhstan," 1.

46. This is in per capita terms. See "Kazakhstan: Living Standards During the Transition," *World Bank Report*, 17520 (March 23, 1998), 4.

47. For a more complete description of this program, see chapter 5 of this volume, "Kazakhstan: The Long-Term Costs of Short-Term Gains," by Pauline Jones Luong.

48. Rosenberg and Saavalainen, "Dealing with Azerbaijan's Oil Boom," 38–40.

49. Rosenberg and Saavalainen, "Dealing with Azerbaijan's Oil Boom," 5.

50. Nonetheless, because of differences in how to divide Caspian oil, it is still not clear who actually owns much of the resources in the Caspian Sea.

51. Charles McClure, "Lessons Learned from the Kazakh Tax Reform: Reduce the Number of Taxes; Resist Incentives for Local Industry," *East/West Executive Guide* 7, no. 4 (April 1997).

52. Daniel Yergin, *The Prize* (New York: Simon and Schuster, 1991); and Robert W. Told, *The Russian Rockefellers* (Stanford, Calif.: Hoover Institution Press, 1976).

53. I am grateful to Laurent Ruseckas for this point.

54. John P. Willerton, *Patronage and Politics in the USSR,* n.d.

55. John Roberts, *Caspian Pipelines* (London: Royal Institute of International Affairs, 1996), 81.

56. In *The Prize*, Yergin describes the strategic importance of Caspian petroleum in both wars. In each case, Germany tried to seize the Caucasian fields to fuel its military machine, but each time attempts to take possession of the fields ultimately failed.

57. Forsythe, "The Politics of Oil in the Caucasus and Central Asia," 10.

58. Some of this shift is due to a long "transition" recession and the devastation caused by the Karabakh conflict, which caused the loss of control over valuable agricultural land and as much as 20 percent of the country's territory. This data, which is the most recent available, is provided from "Country Profile: Azerbaijan 1998/9," *The Economic Intelligence Unit,* 39, from data provided by the State Statistics Committee, Reference Tables 16, 17.

59. This is true for refrigerators, caustic soda, air conditioners, sulphuric acid, pure cotton, footwear, clothes, mineral water, grape wine, champagne, cognac, vegetable oil, butter, and cement, according to the State Statistics Committee, cited in "Country Profile: Azerbaijan," reference table 11.

60. See "Country Profile: Azerbaijan," reference tables 16 and 17, respectively, for these statistics.

61. "A Central Asian Corporate Culture," *Business Week,* October 27, 1997.

62. Karl, *Paradox of Plenty,* 1997.

63. Richard Auty, "Does Kazakhstan Oil Wealth Help or Hinder the Transition?" *Development Discussion Paper,* no. 615 (Harvard Institute for International Development, December 1997).

64. Profits gained by foreign legal entities as a result of their economic activities in Azerbaijan are subject to a 25 percent rate of taxation as compared to the minimum 35 percent rate for domestic entities. Dale Gray, "Energy Tax Reform in Russia and Other Former Soviet Union Countries," *Finance and Development* 35, no. 3 (September 1998).

65. Mark Campbell, Mike Kubena, and Varinder Matharu, "Exploring the Taxation of Oil and Gas," *International Tax Review* 9, no. 7 (July-August 1998): 51–57.

66. Gray notes that the tax burden is generally between 2 and 4 percent for oil-exporters, but in Azerbaijan it is only .79 percent. See his "Energy Tax Reform in Russia and Other Former Soviet Union Countries."

67. Karl, *Paradox of Plenty,* 1997.

68. U.S. Department of State Commercial Guide, 1998

69. He did, however, hold a referendum in 1994 for the purposes of extending his rule. The official results were 1,050,408 in favor and 212 against. See U.S. Department of State Commercial Guide, 1998.

70. Alec Rasizade, "Azerbaijan and the Oil Trade: Prospects and Pitfalls," *Brown Review of World Affairs* 4, no. 2 (Summer–Fall 1997).

71. Campbell, et al., "Exploring the Taxation of Oil and Gas," 51–57.

72. Roger Thomas, British Ambassador to Azerbaijan, made this statement when commenting on the prospects for political stability. See Arab Press Service Organiza-

tion, July 13, 1998. *The Economist* reports that government positions are generally bought, with the job of tax collector, for example, selling for about $50,000. Also see "Foreign Investment in Azerbaijan's Mineral Resource Sector," *East/West Executive Guide,* July 1, 1996; and chapter 4, "Azerbaijan: The Politicization of Oil," by David Hoffman, in this volume.

73. Given this incentive structure, most rulers will act the same way. President Elchibey, for example, Aliyev's predecessor, was reportedly visited by the former UK Prime Minister Margaret Thatcher in September 1992, who deposited a signature bonus from British Petroleum for several million dollars on his desk. According to one official, this contribution helped persuade Elchibey to grant the company the exclusive right to develop the Chirag oil field. Sampson, "Lubricating the Caspian."

74. Information on these practices is not easy to find. On corruption and nepotism, see "Survey of Central Asia," *The Economist,* February 7, 1998. On the tax laws, see Lisa Gialdini, "Tax Code Amendments to the Kazak Tax Code," *East/West Executive Guide* 7, no. 2 (February 1, 1997); and "Kazakhstan: 1998 Investment Climate Statement," June 30, 1998.

75. See Hugh Pope, "Autocracy is Spreading in Former Soviet States," *Wall Street Journal,* October 14, 1998. Also see Sampson, "Lubricating the Caspian." The Economist Intelligence Unit (1998), and *The New York Times,* October 4, 1998.

76. *RFE/RL,* March 16, 1999.

77. See, for example, Roberts, *Caspian Pipelines.*

4

Azerbaijan: The Politicization of Oil

David I. Hoffman

Forty kilometers into the Caspian, *Neft Dashlari*, or Oily Rocks as it is known to its foreign guests, presents a surreal, almost macabre sight. Sitting on stilts off the coast of Azerbaijan, the complex is a maze of oil rigs, eight-story apartment blocks, steel pipes, and movie theaters, all linked together by over 125 miles of roads. For a moment, one can almost imagine the awe and satisfaction of Soviet engineers as they neared completion of their creation in 1947—a massive industrial complex located in the middle of the sea; a triumph of Soviet engineering and planning!

The illusion, however, is fleeting. It is with good reason that almost none of *Neft Dashlari's* foreign visitors stay overnight. Rusting and decrepit, the facility is literally falling into the sea, with miles of roads already submerged beneath the oily soup that is the Caspian. Workers assigned to the field, in a quest for higher and drier accommodations, have taken to converting a portion of their wages into bribes to secure housing on the third floor or higher. Around some dormitories, the waterline now stands at the second-floor windows.

Although fully one-third of Oily Rocks's 600 wells are inoperative or inaccessible, there is no thought here of discontinuing operations, or of pausing for even a second for a thorough retrofit, regrade, and rethinking of the facility's mission. The site, despite its imperfections, still produces more than half of the Republic of Azerbaijan's total crude oil output, and therein lies the problem. For a young, revenue-starved state coupled to the remnants of an aging, inefficient oil sector, there can be no other direction than forward when it comes to questions of energy development. And so operations at Oily Rocks continue, as they have since 1947, extracting oil from the shallow water portion of the Absheron geological trend. At the same time, the Azerbaijani government, to the extent that it has proven possible, has striven to revive Oily Rocks with foreign investment, which has resulted in several shiny, anomalous additions grafted onto the sinking hulk.

More than any other single metaphor, Oily Rocks symbolizes the role played by the oil industry in the Republic of Azerbaijan's young, turbulent history. Dilapidated and aged, and befouling the waters around it, the oil extraction facility nevertheless is, like the energy sector in general, an absolutely critical piece of the Azerbaijani budget picture. Meanwhile, for-eign-funded equipment, such as the bright yellow Pennzoil-built gas com-pression plant, coexist uneasily with half-submerged living quarters and leak-ing pipes, mirroring the difficulties encountered in integrating western capital with the Azerbaijani government's almost-religious faith in the energy sector as the country's primary magnet for foreign investment.

The energy sector's importance to Azerbaijan goes far beyond economics, in ways not immediately evident 40 kilometers into the Caspian. At different times, for different people, oil has answered to a spectrum of names in the Azerbaijani mindset, from that of "black gold paving the road to Kuwait," to a cursed commodity, controlled by and benefiting but a few, to the "hydro-carbon weapon" that would ultimately guarantee Nagorno-Karabakh's return to Azerbaijani control. Beyond the public discourse, however, energy—and in particular, the oil industry—is central to the Azerbaijani economy as the likely engine of any future economic development. It also anchors Azerbaijan's foreign policy. And, most quietly and perhaps most importantly, the country's energy sector is playing a fundamental role in the process of state-building in the post-Soviet era.

The road to a rosy hydrocarbon future, however, is strewn with pitfalls: Dutch Disease, foreign policies held hostage by the vagaries of the world oil market, and malformed states barely capable of discharging the most basic of duties are all too common in the pantheon of the world's petro-states.[1] And thus we arrive once again at *Neft Dashlari:* for the political leadership in Baku, the focus on exploiting the country's geological treasures is worth literally any cost, including the ironic possibility that this undeniably rich bounty may ultimately prove a bitter pill to hopes for long-term economic growth and political stability.

BIRTHING PAINS AND MOTHER'S MILK

To understand why oil-led development has been an unquestioned imperative for Azerbaijan, rather than an option for sober consideration, one must first turn to the circumstances surrounding the country's birth. Three fundamen-tal facts mandate Azerbaijan's dependence on the development and expansion of its oil industry. The first is economic: since severing ties with the then-disintegrating Soviet Union in 1991, Azerbaijan has had to confront the real-ity that, aside from the oil industry, no other viable economic sectors exist that are capable of generating income and foreign exchange in substantial volumes.

Since 1991, the Azerbaijani economy has been rife with *potential*, but when faced with immediate budget shortfalls, domestic political instability, a moribund industrial base, and an escalating war, Azerbaijan could ill afford the "ramping-up" period necessary for capital investment and restructuring to catalyze other latent sectors of its economy.

Furthermore, having been tied closely to the budgetary and trading patterns of the old Soviet centralized economy, Azerbaijan found itself particularly vulnerable to the disruption of trade links with its neighbors. Russia's unilateral closure of its border with Azerbaijan in 1994 (ostensibly in response to the conflict in Chechnya) exacerbated this further.[2] To date, despite the rosy statistical picture presented by the government and the IMF,[3] the fact remains that Azerbaijan's official economy is largely barren: its industry at a virtual standstill, the chemical and petrochemical sectors (the former workhorses of the republic's economy) dead or dying, and once-vibrant manufacturing and export sectors laid low by severed trade links[4] and a strong national currency.[5] The Azerbaijani economy in its current state is incapable of providing sufficient income to the national budget due to problems with tax collection, regulatory enforcement, and a malleable and opaque legal regime. This inability to generate revenue has led to a realization at the highest levels of government that remittances from oil contracts, in the form of direct investment, bonus payments, and oil sales, currently present the best hope for securing cash flow for the country. Put bluntly, oil has been the only viable, export-capable economic lever independent Azerbaijan has known.

Azerbaijan's dependence on oil is also firmly rooted in its geographical and geopolitical isolation. Sandwiched between its enemy Armenia to the west, an increasingly centripetal-minded and openly pro-Armenia Russia to the north, and an Iran with whom relations have been chilly at best to the south, Azerbaijan began life as an independent country with precious little currency in the international relations game.[6] Its two closest allies—Turkey and the United States—meanwhile, have at different times expressed their "enthusiasm" for Azerbaijan by sponsoring a coup (1995's Turkish-supported OPON rebellion) and passing Section 907 of the 1992 Freedom Support Act, respectively.[7] The latter singled-out Azerbaijan as the only Soviet successor state not eligible for direct government-to-government assistance from the United States.[8]

Given Azerbaijan's difficult position, then, it is little surprise that the government in Baku saw its oil industry as perhaps the only available asset capable of forging closer ties with foreign states. Oil, however, seldom if ever translates automatically into newfound friends on the international stage. In the case of Azerbaijan, however, an appropriate vehicle was found in the form of foreign energy companies, in particular western oil companies, whose long-term, capital-intensive presence in Azerbaijan would, it was thought, bend their home government towards more sympathetic relations with

Azerbaijan. After all, between 1991 and 1994, the country was suffering regular humiliations not only on the battlefields of Nagorno-Karabakh, but in the war for international sympathy, where it could muster no influential response to the powerful lobby of the Armenian diaspora in the West. Thus, from its inception, Azerbaijan has been driven to open its oil resources to foreign exploitation as a lure for attracting strategically significant foreign investors, and ultimately converting these relationships into diplomatic currency.[9] The formation of pro-Azerbaijan lobbying groups in the West, such as the Caspian Business Group, testifies to the partial success of these efforts.

That oil plays a crucial geopolitical role for Azerbaijan is further reflected in the sequencing of the government's development of its energy resources. Since 1992, Azerbaijan's strategy for developing the country's oil resources has prioritized the development of its offshore fields.[10] This policy was formulated and is maintained at the top levels of the political establishment (specifically, within the presidential apparatus), is executed by the State Oil Company of the Azerbaijani Republic (SOCAR), and has been remarkably consistent in the face of repeated regime changes since independence.[11] Unlike onshore exploration and development contracts (at least until 1998), agreements covering the Azerbaijani offshore were framed as production-sharing agreements (PSA), a formula that boasts one overwhelming benefit over other investment-recovery schemes: PSA contracts, once signed by both parties, are ratified by the Azerbaijani parliament (*Milli Majlis*), and thus assume the force of law. This provides increased levels of contract stability, commensurate with the PSA's insulation from notoriously capricious Azerbaijani domestic legislation. As opposed to standard tax-and-royalty schemes, production-sharing agreements provide a physical mechanism for rendering to the Azerbaijani state its share of profits, while allowing foreign energy companies to recoup their investments. Foreign participants in Azerbaijani PSAs recover their capital and operating costs in the form of a share of crude production at the beginning of the production cycle. The remainder of a field's oil output is then split between the state and its foreign partner(s) according to a formula agreed upon for each individual PSA.

Compared to onshore prospects, oil projects in the Caspian Sea are perceived by the Azerbaijani leadership as having critical geopolitical importance.[12] This perception, combined with internal political instability in Azerbaijan and smoldering disputes over the legal status of the Caspian Sea, dictated a dramatic and early gesture by the Azerbaijani government in its development of offshore oil reserves. In 1994, when the "contract of the century" was signed with the Azerbaijan International Operating Company (AIOC), the Azerbaijani state was notoriously unstable, reeling from hyperinflation at home, wholesale collapse on the Karabakh battlefield, and the prospect of increasing Russian and Iranian pressure on its strategic flanks.[13] After barely a year in power, President Heydar Aliyev desperately needed to consolidate and secure his personal position. Having himself been witness to four coups d'état in Baku over the previous four years (the last of which he rode into the president's office), Aliyev was extremely keen to secure both strong international allies to fortify himself against internal and external enemies, and a ready source of hard currency for the reeling Azerbaijani economy. The signing of massive, multi-decade offshore contracts with the BP-Amoco-led AIOC consortium and other foreign oil companies helped to intertwine foreign commercial interests with the country's fate, while simultaneously pulling major western governments into a closer strategic embrace with Azerbaijan. As for cash flow, offshore PSAs, due to their scope and breadth, promised not only massive long-term hard currency revenues for the Azerbaijani economy, but substantial and immediate bonus payments for the government.[14]

Azerbaijan's anemic economy and international isolation were clearly important birthing pains that dictated an early and unreflecting embrace of oil development. To these two factors, however, a third must be added: Nagorno-Karabakh. Before Azerbaijan's appearance as an independent country, even before its first unsure steps towards state-building in the late 1980s, the conflict in Nagorno-Karabakh raged. The struggle for the ethnically Armenian yet juridically Azerbaijani territory would eventually pit Azerbaijan not only against secessionists in Karabakh, but the government of neighboring Armenia and the weight of the Russian foreign policy establishment.[15]

The dissolution of the Soviet Union saw the conflict over Nagorno-Karabakh grow in scale from neighborhood violence and homemade weaponry to de facto interstate conflict punctuated by the use of heavy weaponry and air power. Azerbaijan's loss of lives and territory accelerated at a corresponding rate, reflecting both domestic political turmoil and the absence of a trained, well-equipped national army. By the time a leaky ceasefire was arranged in mid-1994, 20 percent of Azerbaijan's pre-war territory had been wrested from Baku's control, and over 800,000 of the country's approximately seven million citizens had been dislocated from their homes.[16] The loss of Nagorno-Karabakh and adjoining regions of Azerbaijan proper produced a shock that reverberates to this day for Azeris of all social strata. Direct economic consequences have been massive, as the occupied region represents some of Azerbaijan's most productive agricultural areas. Politically, meanwhile, the rupture of Azerbaijan's territorial integrity has weakened the authority of the central government and affected its ability to exert full control over other would-be autonomous ethnic groups, such as the Lezgins in northern Azerbaijan.[17] Finally, the presence of close to a million internal refugees ("internally displaced persons") in a country of only seven million people has resulted in extreme social strains, with cultural and economic differences between refugees and "locals" increasingly breeding mutual resentment between the two communities in a pattern repeated throughout Azerbaijan.

The fighting for Nagorno-Karabakh was a savage and ultimately futile affair for Azerbaijan. It was also expensive, and put enormous strain on Azerbaijan's state finances. In addition to the financial, material, human, and geographical losses incurred from the war, the state treasury found it necessary to compensate for a "reform gap," which resulted from an absence of foreign and domestic investment and economic restructuring that might have been, were it not for the war. Until the 1994 cease-fire, war losses, the failure to invest in or reform the economy, and the lack of time or occasion to craft long-term strategies for economic development and fiscal management placed oil development at the forefront of the government's efforts to generate revenue. Oil thus substituted for the constructive capital investments not being made by Azerbaijan itself or almost anyone else at the time.

OIL AND POLITICS IN AZERBAIJAN: THE LAY OF THE LAND

Given the importance of oil for Azerbaijan, it is no surprise that the country's political structures and oil sector are intimately linked. Formally, virtually all aspects of the Azerbaijani oil sector, including relations with foreign companies, are the prerogative of the State Oil Company of the Azerbaijani Republic. SOCAR is not only responsible for negotiating and implementing oil sector projects, but it is also charged with international oil marketing. In principle, SOCAR is governed by a board of directors that consists of 13 individual and nine corporate permanent members; in reality, however, three principal actors direct most of SOCAR's affairs: SOCAR President Natik Aliyev (no relation to Azerbaijan's President Heydar Aliyev) SOCAR Vice President (and son of Azerbaijan's president) Ilham Aliyev, and Valekh Aleskerov, who conducts SOCAR's negotiations with foreign oil companies. The president of SOCAR simultaneously carries the rank of minister, thus reporting to the prime minister. This official chain of command is, however, overshadowed by a parallel network of informal and kinship ties between SOCAR and the executive branch, the most prominent of which is the relationship between Azerbaijan's head of state and SOCAR's vice president. Top-level strategic decisions affecting the direction of Azerbaijani hydrocarbon development are made by and transmitted through this network, rather than within the formal government hierarchy. This arrangement allows President Aliyev to control the country's oil industry, and precludes any political division of control over energy resources with potential rivals.

The ties that bind the Azerbaijani oil industry to the president's office also condition SOCAR's relations with other state bodies. The priority accorded to the energy sector has given SOCAR near-exclusive control over the development of the country's oil industry. Set against the backdrop of a feeble state defined by weak, politically malleable institutions, this has led to the virtual absence of the administrative turf wars so frequent in other post-communist states. Potentially powerful state bodies, notably those involved in investment activities, privatization, or regulatory duties, have proven incapable of either diminishing SOCAR's hold over the oil industry or of extending their influence into this lucrative sector. Revealingly, SOCAR has also proven relatively resistant to the administrative restructurings, leadership changes, and shifting mandates that have characterized other state bodies.

Azerbaijan's political order is dominated by the figure of President Heydar Aliyev. Aliyev rode a wave of political chaos, social anarchy, and military insurrection to return to power in 1993. He has since consolidated his position through the skillful manipulation of state institutions and formal political structures. This process has succeeded in strengthening Aliyev's position, but has also fundamentally undermined the process of state-building. The Azerbaijani constitution, adopted in a November 1995 referendum, established a system of government based on a nominal division of powers between a strong presidency, a legislature with the power to approve the budget and impeach the president, and a judiciary with limited independence. In reality, however, these state institutions have been deliberately engineered to reinforce, rather than moderate, the power of the executive. The three nominally independent high courts—the Constitutional Court, Supreme Court, and High Economic Court—are extremely susceptible to executive influence, and moreover are staffed primarily with judges beholden to Aliyev. The Parliament, meanwhile, exercises virtually no legislative initiative independent of the executive. Following "flawed"[18] parliamentary elections in 1995, a total of eight political parties gained representation in the *Milli Majlis.* Of the 125 seats, only eight were occupied by the political opposition, the rest going to President Aliyev's *Yeni Azerbaycan* party, allied pro-government parties, and primarily pro-government "unaffiliated" candidates.[19]

Unlike his predecessors, Heydar Aliyev has had the time and political breathing space to consolidate and considerably strengthen his position following his ascension to power. As was graphically demonstrated by his suppression of Surat Husseinov's 1994 coup attempt, Aliyev is no longer beholden to any one of the forces that ushered him to power in 1993.[20] Rather, his position is now not only insulated by overlapping formal state institutions, but (following two imperfect if on-schedule presidential elections in October 1993 and 1998) wrapped in a veneer of nominal electoral legitimacy. The latter has proven especially useful in normalizing relations with international financial institutions and western governments.

At first glance, Azerbaijani politics are demarcated, as in many countries, by state institutions: specific agencies and organs of the state that are nominally oriented around functional tasks, and staffed by bureaucrats who are charged with executing highly delineated duties and selected and promoted according to a standardized, usually meritocratic rationale. A deeper analysis, however, reveals an alternative and ultimately more influential logic at work: more than political institutions, or even laws, what determines individuals' or organizations' political (and often economic) worth is their proximity to President Aliyev.[21] A cordial relationship with Azerbaijan's chief executive is the ultimate political capital in the country, and pays both economic and political dividends. Politically, extensive patronage networks dominate the various government ministries and judiciary, and provide powerful yet informal avenues for advancement and promotion. Economically, a close relationship with President Aliyev or his family can translate into favorable terms for business in the form of preferential contracts, tax exemptions, relaxed regulation, and access to profitable state-owned assets. Azerbaijan's politics are thus best thought of *not* as a series of impartial, parallel state bodies and official institutions, but rather as a wheel, where practically all significant spokes radiate out from the president's office.

The highest levels of the Azerbaijani political system are dominated by a series of tight networks, centered on President Aliyev, which serve to reinforce economic and political interests of the ruling elite. Of these, the most prominent is clearly the regional "tribe" composed of Azeris from Armenia (*Yeraz*) and the Azerbaijani enclave of Nakhichevan. President Aliyev himself is from Nakhichevan, and most of his inner circle hail from this regionally defined "tribe." Another important network consists of bureaucrats and "new businessmen" who were Soviet-era associates of the president. Much of Aliyev's political power in the earliest part of his presidency stemmed directly from strong and persisting ties forged during the Soviet era.

Although Nakhichevani and *Yeraz,* along with Soviet-era associations, constitute the inner circle of political power in Azerbaijan, the direct guarantor of President Aliyev's (and, hence, these networks') political power ultimately stems from his control over the "power ministries": the Army, the Ministry of Internal Affairs, and the Ministry of National Security (successor to the Soviet KGB). These institutions protect the president from overt coup attempts and manipulate public political discourse in the president's favor through media censorship, suppression of political dissidents, and other means.[22] The top levels of the power ministries themselves, in turn, are disproportionately staffed with Azeris from Aliyev's regional "tribe," and also granted extensive leeway in pursuing profitable economic ventures, thus ensuring their loyalty.[23] In short, the linking of economic and political networks in Azerbaijan provides incentives for the country's elite to support the president.

TWO SIDES OF THE SAME COIN?
THE OIL-POLITICS DYNAMIC

In Azerbaijan, the relationship between politics and the evolution of the oil industry is not static. These two nominally independent worlds not only reflect a common logic of centralization of control and weak institutionalization, but interact with each other in significant ways that shape both the way business is done in the oil industry and the course of state-building in Azerbaijan. As illustrated in the previous section, the mechanisms for controlling Azerbaijan's oil sector are tightly clustered around the president. As the politically acceptable agent of the government, SOCAR maintains a near-monopoly on the management of the country's oil industry. Although SOCAR contracts must be ratified by the Parliament, this is a mere formality, not only due to the tight ties between SOCAR and President Aliyev, but also to the feebleness of Parliament itself. In its nearly six years of existence, it has not once rejected or even returned for review an oil contract put before it. To some extent, this arrangement has a salutary effect on the oil industry, helping to channel foreign investment into the country. Foreign energy companies operating in Azerbaijan have reported that SOCAR's strong position and proximity to the seat of political power have established it as a de facto "one-stop shopping" contact for foreign investors. The overriding political favoritism undergirding SOCAR's position also minimizes bureaucratic interference at the higher levels of the government, thus helping oil projects sidestep many potential administrative pitfalls and delays.

The absence of strong, relatively autonomous state institutions in Azerbaijan has proven something of a mixed blessing for energy projects in the country. Oil contracts tend to operate outside normal regulatory channels. Negotiated by SOCAR's top leadership and implemented by the company, contracts assume the force of law once ratified. Parliament, irrespective of the constitution, exercises virtually no oversight powers, and there is a distinct lack of a horizontal distribution of state power within the government, resulting in a heavily deformed regulatory and legal regime.[24] While this may prove helpful to foreign oil companies in some instances (environmental and labor laws, for example, can prove elastic), it is also a situation rife with danger. Weakly institutionalized laws and regulations make for a capricious legal regime that can and does expose foreign companies to irregular legal, regulatory, and tax scrutiny. Foreign businessmen operating in Azerbaijan have complained of being taxed "at the pleasure" of the authorities. This trend is further amplified by excessive rent-seeking on the part of state bureaucrats, who earn ridiculously low official wages, to seek extra-budgetary income—a direct manifestation of the lack of progress in developing professional or regularized state institutions.[25]

Owing to SOCAR's relative autonomy, close presidential ties, and lack of legislative oversight, there is no fixed method for negotiating oil contracts in Azerbaijan. This has led to conflicting interpretations of contract terms between the government and foreign oil companies. The Azerbaijani side has, on multiple occasions, proven willing to unilaterally interpret agreements and laws to its own advantage, underscoring the lack of an institutional basis for contract enforcement. Perhaps the most visible expression of contractual ambiguity came in 1998, when it was discovered that the second pipeline for AIOC "early oil" (this one running west, through Georgia, to the Black Sea port at Supsa) would cost $591 million, and not $315 million as originally estimated and budgeted.[26] Although SOCAR had approved the original budget, this did not represent an AIOC commitment to deliver at that cost. Nevertheless, AIOC's desire to cover cost overruns with oil sales revenue was rejected out of hand by SOCAR, which claimed that the additional $276 million should be paid for by AIOC—a claim with no grounding in the relevant PSA.[27]

The political realities of Azerbaijan affect oil contracts not only when contracts are implemented, but also when they are formulated. Since 1994, for example, oil contracts signed between SOCAR and major foreign oil companies have increasingly favored the Azerbaijani side. At SOCAR's insistence, its participation in each of the 17 PSAs has steadily risen, while its up-front expenses are carried by other consortia members. For example, the AIOC, Caspian International Petroleum Company (CIPCO), and Shah-Deniz PSAs provided SOCAR with stakes worth 10, 7.5, and 10 percent stake respectively. SOCAR now regularly demands a 50 percent stake in each new project, even through the dramatic slump in world oil prices in 1998 and 1999.

Other project parameters, such as foreign subcontractors' taxes, value-added tax terms, and bonus payments, have also proven unpredictable as a result of the ad hoc manner in which contracts are negotiated. The absence of a standardized template for devising, implementing, and regulating energy contracts has also allowed SOCAR to alter the cost-recovery and profit-sharing schedule curve to its own advantage. Generally, within the PSA framework foreign consortium members are allowed to recover all costs (capital expenses and investments) up front, at the early stages of production. Only then is "profit oil" slated to begin, split unevenly between the Azerbaijani government (commanding the lion's share) and consortium members. In 1998, however, Houston-based Frontera Resources found that SOCAR's terms had tightened somewhat. In negotiating a PSA for the rights to the Kyursangi and Karabagly onshore fields in western Azerbaijan, Frontera confronted SOCAR's insistence that it flatten its production-sharing curve, spreading out the cost recovery period and allowing for early profit oil. Given the Azerbaijani government's strategic imperative to increase revenue inflows across the board, this move made perfect sense.

The interaction between the political superstructure and the oil sector in Azerbaijan is not confined to the influence of politics on oil deals; causality very much runs in the other direction as well. Oil helps to shape the political arena in Azerbaijan in two critical ways: by expediting the geographical centralization of power, and as a potentially potent issue in the Azerbaijani domestic political realm. The geography of oil in Azerbaijan plays an important role. Unlike nearby Kazakhstan, where the richest oil deposits lie over 1,000 kilometers from the country's political and financial centers, Azerbaijan's oil fields are concentrated heavily in and around the Absheron peninsula, where Baku, the seat of political power, sits. In a country where oil has proven the only resource of appreciable, recoverable value, this fact of geography is inherently political, since it precludes any resource-based bargaining between the political center and other regions of the country. In conjunction with an unrepresentative political regime distinguished by the informal, regionally based *concentration* of power, rather than a formal, regional *distribution* of power, this has led to a situation where other regions of the country suffer greatly.[28] Absent both a voice in the country's regionally exclusive ruling elite, as well as the strategic resources to leverage support from the center, other regions of Azerbaijan (such as Gyandja, Azerbaijan's second-largest city) have gone years without even the most basic of government services, such as natural gas supplies or road repair.

Energy resources in Azerbaijan thus serve to reinforce central authority, instead of playing the role of a "bargaining chip" between center and regions. In this capacity, oil contributes to the zero-sum nature of Azerbaijani domestic politics and the stubbornness of government-opposition conflict. It also raises the ominous spectre of further ethnic secessionist movements in Lezgin- and Talish-populated regions in the northeast and south, respectively.[29] With no resource-driven incentive to share political power with other regional or political affiliations, the authorities have perhaps unwittingly tipped the logic of their domestic opponents away from one of sharing resources and cooperating with the ruling elite to seizing and replacing both resources and political opponents simultaneously.

The other primary guise in which Azerbaijan's oil sector would be expected to act within the political arena is the latent issue of pent-up frustrations. Given that energy development and foreign investment represent by far the largest direct foreign presence in Azerbaijan and constitute the pillar of the government's plans for economic development, it seems altogether logical that oil would gain momentum as a political lightning rod for frustrations related to the corruption and low living standards that have permeated the country. In fact, ex-President and Azerbaijan Popular Front (AzPF) Chairman Abulfaz Elchibey, upon returning to Baku from internal exile in 1997, explicitly attacked the involvement of foreign energy companies in the development of Azerbaijan's oil industry. In particular, Elchibey singled out the AIOC pipeline project for criticism, at one point going so far as to hint that AzPF partisans in the Gazakh and Agstafa regions might take military action against the pipeline if the project did not "serve the interests of the Azerbaijani people."[30] Understandably, these statements proved a cause for concern for western investors, as they exposed the possibility of using oil as a weapon for attacking the government, and thus, by extension, its foreign partners.

Given the stature and potency of oil as a potential target, it is revealing that the "foreign oil card" has thus far *not* emerged as an explicit campaign issue in Azerbaijani politics. The reasons for this apparent anomaly are informative, and speak to the sensitivity of the country's geopolitical situation. Without a doubt, recent years have seen a rise in popular disenchantment with the presence of foreign oil companies, and especially the AIOC. This is reflected in the increasingly critical press, as well as on the streets of Baku, where almost all Azerbaijanis can identify AIOC's gated headquarters and recite the oft-cited figure of $40 billion in oil contracts already secured with foreign companies.[31] Juxtaposed against the persistently feeble, yet increasingly uneven, standards of living in the country, this would seem to portend wide-scale popular resentment against foreign exploitation of Azerbaijani resources. Indeed, beginning in 1998 the disparity between wages paid to foreigners and to local Azerbaijanis by foreign firms began to receive attention by the Baku print media.

Nevertheless, the initial euphoria felt by many Azerbaijanis four or five years ago at the onset of the oil "boom" has been followed primarily by resignation, and *not* politically directed rage. The population of Baku, in particular, is already largely inured against charges of government corruption and continuing Soviet-style inflation of economic indicators, and is thus less likely to seize upon the opposition's charges of opaque, possibly illegal oil negotiations as a commanding theme during national elections. This stems primarily from the strong public disillusionment with the main opposition parties (the AzPF and Yeni Musavat), whose tenure in power (1992–93) was marked by economic chaos, wartime defeats, rampant corruption, and midnight gangs roaming the streets looking for draft dodgers.

Furthermore, the state-owned and state-influenced media have actively— and largely successfully—attenuated oil's public potency by redirecting public attention elsewhere after the earlier stream of hyperbolic references to "petro-billions" began to reflect poorly on the unpleasant reality of the economic lot of the average Azerbaijani citizen. Attacks on the oil industry have faded even from the pages of newspapers openly sympathetic to the opposition, such as the AzPF-oriented *Azadlyg*. This tendency seems to reflect the understanding that a general disillusionment with oil as "the way to Kuwait" is unlikely to evolve into a general groundswell of public opinion against foreign-aided hydrocarbon development, primarily because it is politically difficult to articulate a coherent alternative to foreign (and especially western) involvement in the oil sector. Thus, while dashed expectations and an increasingly cynical attitude towards "oil riches" will continue to be the cause for much grumbling and complaining, there is at the same time an extremely broad consensus that oil development remains Azerbaijan's best hope for consolidating its independence from Russia, as well as an understanding that Azerbaijan is incapable of developing, much less exporting, these resources on its own. The distributive inequality characteristic of Azerbaijani oil development, while a potent *potential* issue, has thus far largely withered on the vine for lack of a credible opposition to exploit it.[32]

The *failure* of oil (or of oil-related corruption and disappointment) to resonate within the Azerbaijani public as an election issue also speaks to the importance of Nagorno-Karabakh for both the population and the country's political leaders. Azerbaijan's experience with Nagorno-Karabakh continues to overshadow oil, both publicly and privately, as an emotional issue. Even at the apogee of hyperbole, following the much-ballyhooed 1994 signing of the AIOC "contract of the century," this enthusiasm was tempered by the understanding (or at least suspicion) that converting natural resources into national wealth would take time. Even as oil wealth has failed to raise living standards appreciably for most Azerbaijanis, there is an understanding today that international political momentum is such that sooner or later large volumes of oil will be exported from Azerbaijan, and that some monies must inevitably reach the population. In contrast, with the spring 1998 ouster of Armenian President Ter-Petrossian, the prospect of a return of the occupied territories to Azerbaijan seems more distant than ever, if indeed it can be imagined at all.[33] Unlike long-term macroeconomic stability or international oil contracts, the liberation of territory is a tangible, non-fungible issue, one with more direct symbolic and personal meaning for most Azerbaijanis.

Nevertheless, oil remains a rallying point for domestic critics. If President Aliyev fails to generate even the appearance of progress at the OSCE-mediated talks on Karabakh—or worse, if military action resumes and Azerbaijan fares badly on the battlefield—then this will open up the door to wider criticism of the government in Baku. In conditions of relative domestic stability and quiet along the frontlines, the oil issue is not likely to prove an effective weapon capable of stinging President Aliyev. Failure—real or perceived—on the Karabakh issue, however, would weaken the current regime immeasurably, and expose its oil dealings to broad and effective criticism.

"BLACK GOLD" OR "THE DEVIL'S EXCREMENT"?

Will energy revenues bring a better life to the citizens of Azerbaijan? With the rich heydays of Arab Gulf states in mind, one is tempted by the description of oil as "black gold." A different image, however, is provided by OPEC founder Juan Pablo Perez Alfonzo, who once lamented that oil, "is the devil's excrement. . . . We are drowning in the devil's excrement."[34] Of course, oil in and of itself is neither of these—what ultimately determines the social, political, and economic impact of energy resources on a given country are the revenues generated, be they through sales of oil volumes, oil-related foreign investment, or bonus payments. In time, the net effect of Azerbaijan's "oil wealth" will vary according to the ways in which such capital inflows are deployed within the domestic economy—patterns that are themselves dependent on the institutional and societal mechanisms active in channeling, dividing, and sub-dividing revenue.

In the case of Azerbaijan, it is still early to speak authoritatively of the long-term legacy of hydrocarbon-derived revenues on the state and society. Although oil-related revenues already accounted for nearly half (47 percent) of the consolidated budget in 1997, the "wall of money" anticipated by some pundits will not materialize for several years at the earliest.[35] Revenue flows of this magnitude will begin to emerge only as large-scale exploration and production projects begin producing results early in the next decade, coinciding with the construction of one or more large-capacity export pipelines. The latter represents the clearing of an important export bottleneck, and will have a multiplier effect on Azerbaijani upstream development (assuming new offshore oil fields are discovered). The Azerbaijani people and state will be constrained to wait for the bulk of their oil revenues due to the structuring of PSAs. At 1997 prices, the AIOC, for example, was expected to generate approximately $2.5 billion per year at peak output and after subtracting costs.[36] This revenue is to be split between the Azerbaijani government and the AIOC shareholders. And, although SOCAR and the government are slated to realize approximately 80 percent of total revenues, the production-sharing formula dictates that Azerbaijan will only begin to recover its lion's share after AIOC's expenses and investments have been recouped—a benchmark that, depending on the price of oil, may not be reached until the end of the next decade.

Although the full impact of "big" oil wealth is still years away for Azerbaijan, certain institutional and political indicators allow for a glimpse of the likely long-term consequences of oil-led economic development. Indeed, although oil money at peak volumes is not yet flowing into Azerbaijan, many of the factors that will play a role in channeling and redirecting these monies are present or coalescing today. Clearly, SOCAR is one such institutional bellwether. Although reform and possible reorganization (in conjunction with the proposed formation of a Ministry of Fuel and Energy) are likely events in SOCAR's future, both the company's basic role and its powers are unlikely to change as long as the current regime or a sympathetic successor is in power.[37] SOCAR is likely to remain the primary mediator between the Azerbaijani government and foreign oil companies. It will also continue to be SOCAR's prerogative (under the close supervision of the executive) to remit the resulting revenues to the state's coffers. The path along which oil monies are "digested" by the Azerbaijani public and private economies is likely to remain opaque due to the continued absence of parliamentary oversight over SOCAR and Parliament's minimal role in national budget formation.

The level of state-building in Azerbaijan, and the nature of the political system provide two further barometers of oil revenues' likely impact. Without rationalized, task-oriented institutions capable of discharging the regulatory, extractive, and redistributive duties of the state, it is highly unlikely that the effects of large-volume oil monies will extend beyond a small, politically and regionally connected circle of the Azerbaijani population. With state institutions deformed in order to insulate presidential power, the Azerbaijani government has displayed little inclination or ability to develop the mechanisms necessary for channeling oil wealth into the national economy. Impartial budgetary controls, long-term economic planning, and bureaucratic reform (especially the implementation of a test-based, meritocratic system of recruitment and advancement, combined with higher official wages and institutional ethos) are either wholly lacking or politicized to near-ineffectiveness. As a result, it is unlikely that the Azerbaijani government will be capable of converting oil revenues into republic-wide economic growth. Indeed, without fully-funded, dependable, and apolitical apparatuses to generate statistics and collect taxes, the Azerbaijani state is hard-pressed even to stay informed of the current state of its own economy, much less accurately plan for the rational deployment of future oil revenues.

The nature of the Azerbaijani political system adds to this problem. In many ways, politics in Azerbaijan describes a system maintained almost entirely by rent-seeking. Since it is generally impossible for bureaucrats to earn a living from their official wages, graft has become an imperative of governing the country, from the lowliest customs officials to a state minister being paid an official salary of $100 or less per month.[38] State officials, in this schema, often depend on their positions to provide access to rent-seeking opportunities.[39] Moreover, due to the lack of neutral and universal civil service testing, bureaucratic recruitment and advancement are punctuated by personalism and political loyalty.[40] This suggests an extremely close relationship between the ruling regime and state institutions, and further contributes to the zero-sum nature of regime-opposition struggle. As argued previously, rather than power-sharing, the dominant idiom thus becomes regime change—implying an attendant "cleansing" of state institutions through purges or de-politicization of bureaucrats.

As a result, the arrival of a "wall of money" from oil revenues is unlikely to diminish or eradicate rent-seeking within the political system. Rather, much of this oil wealth will be spent by the regime to buy and continue to maintain the political support of strategically critical subsets of the Azerbaijani population, most notably the *Yeraz*/Nakhichevan elite (or whatever regional "clan" has succeeded them). Oil-derived capital flows are thus also likely to be diverted to covering up budget revenue holes caused by persistently feeble tax collection, leaky regulation, and other consequences of less-than-professional state administration.

The current path of state and regime development in Azerbaijan also points to a strong susceptibility to Dutch Disease. This occurs when disproportionate investment into a specific extractive industry (oil, in this case) causes wage and price distress in other sectors, ultimately leading to the distorted growth of services, transportation, and other non-tradables. Agriculture and non-oil industries subsequently suffer due to the resulting overvaluation of the national currency.[41] Foreign investment in Azerbaijan, which comprises 68 percent of total investment, is overwhelmingly concentrated (74 percent) in the oil sector. Many other preconditions of Dutch Disease are evident in Azerbaijan, including: (1) the strong dependence on a single natural resource; (2) an evaporating industrial and agricultural base; (3) the preponderance of state ownership in the economy and in total enterprise debt; (4) an autocratic political process with no clear succession mechanism; (5) the concentration of political power in a single individual; and (6) time-inconsistent spending decisions, evincing an inability to articulate a long-term plan for balanced economic development.[42] Budget formation and fiscal planning remain dangerously sensitive to price fluctuations on the world oil market—a fact driven home in 1998 when tax revenues shrank 66.5 percent and royalties to the budget were 60 percent below target in the wake of a dramatic fall in world crude prices.[43]

THE U.S. ROLE: THE NEED FOR ONE VOICE

Set against this tangled backdrop of interdependent energy and state development in Azerbaijan, American foreign policy towards the region has proven only slightly (if at all) less confused. Azerbaijan, from the official standpoint of the U.S. government, is at the same time the pariah excoriated and shunned by Section 907 and the linchpin in the "East-West transportation corridor" that has anchored U.S. geopolitical strategy in Eurasia since 1996. To a large extent, this paradox finds its roots in American domestic politics—a topic outside the scope of this discussion. Nevertheless, without delving into the mists of Washington politics, it is still possible to briefly delineate the relevant American foreign policy interests at hand and to illuminate one or two constructive steps Washington might take in its dealings with Baku.

Since 1996, the U.S. government has espoused a remarkably clear strategy towards the Caspian Sea region of the former USSR. American policy towards Azerbaijan is embedded within this framework, which calls for the development of democratic governments, free and open societies, and functioning market economies in the states of the South Caucasus and Central Asia. Rhetoric aside—its coziness with the almost-universally autocratic regimes of the region has belied the U.S. government's "goal" of democratization—it is possible to identify the core U.S. foreign policy interests in Azerbaijan. These relate primarily to the establishment of a stable export corridor for Caspian Sea energy resources. In the vernacular of "diplo-babble,"[44] this has translated into a strategy based on the somewhat dubious belief that "happiness is multiple pipelines."[45] In reality, the U.S. government is primarily interested in the construction of at least one major east-west energy export route that will be insulated from Russian interference and that will deny Iran the lucrative transportation tariffs associated with such a pipeline.

The desire to promote the export of Azerbaijan's hydrocarbon resources to world markets is rooted in liberal notions of interdependence and positive-sum economics: Azerbaijan's security and ties to the outside world are increased, while the West gains access to further crude supplies. This picture, however, becomes muddled when the interests of the three principal actors involved—the U.S. government, the Azerbaijani government, and the foreign oil companies who are expected to fund and carry out any large transportation projects—are weighed. Washington's push for an east-west transportation axis, ironically, runs counter to the desires of many of the oil companies (American included) active in the region, who balk at the exorbitant costs associated with the vision, and favor an export route through Russia or Iran.[46] Because of sanctions against Iran, however, U.S. policy is to avoid transit routes through the country.[*] This isolation of Iran, moreover, is not necessarily an undesirable outcome from Baku's perspective, but only if it means support for the development of export lines, which the U.S. government has supported with lavish rhetoric but scant funding.[47] These are the contradictions that have paralyzed the main export pipeline routing decision, originally scheduled for October 1998.

In February 1999, President Aliyev's top foreign policy advisor, Vafa Gulazade, created a sensation by suggesting that NATO—in the guise of the United States or Turkey—establish a military base on Azerbaijani territory. Clearly, such a move is ill-advised for the United States, since it would formally end any semblance of Russian-U.S. strategic cooperation and would militarily tie Washington to a state still de facto at war with another nominal American ally, Armenia.[48]

In lieu of military alliances, there are several pragmatic steps that policymakers in Washington can take to deepen and smooth the U.S. relationship with Azerbaijan. The first is to engage the Azerbaijani polity much more broadly than is being done today. Practical steps in this direction would involve developing deeper institution-to-institution ties (in areas other than energy development, such as law enforcement, customs, etc.) as well as contacts with a wider range of political figures in Azerbaijan. Although most political parties in Azerbaijan claim a pro-American worldview, some domestic forces—most prominently the Islamist Party—have begun to reorient towards Iran. Others, such as Zardusht Alizade's Social Democratic Party, and supporters of former President Ayaz Mutalibov, look north to Moscow for inspiration and support. It is also important to decouple U.S. policy from

---

For more on Iran's role as a possible export route for Caspian energy see Chapter 8, "U.S.–Iranian Relations: Competition or Cooperation in the Caspian Sea Basin," by Geoffrey Kemp.

the figure of Heydar Aliyev. Although good relations with Azerbaijan naturally mandate cordiality with President Aliyev, this should not occur at the expense of contacts with other political figures from both the government and opposition camps. As mentioned, given its geographical location, Azerbaijan has many reasons to gravitate towards a Turkish-U.S. axis; establishing contact with the political opposition is unlikely to drive the current regime into the arms of any of its neighbors. Azerbaijan's domestic political situation warrants such a diversification of U.S. policy. President Aliyev's brief medical evacuation to Turkey in January 1999, regardless of whether it was prompted by heart problems or the flu, graphically demonstrated that his departure from the political scene in Azerbaijan will likely *not* follow the script laid down in the constitution, but will instead lead to a naked power grab.[49] Aliyev's unexpected absence ushered in a week of uncertainty and galvanized a spectrum of political forces, including supporters of former President Mutalibov, former speaker of Parliament Rasul Guliyev, and participants in the 1995 OPON uprising. The lesson from this "dress rehearsal" is clear: weak political institutions and the lack of a strong succession mechanism point to a "gloves-off" scenario in the post-Aliyev future, with various mutually exclusive, regionally based political networks locked in a zero-sum struggle for power. In the event that the U.S. government fails to diversify its contacts with other political movements in Azerbaijan, it risks a stillborn relationship with whatever government eventually succeeds President Aliyev. While some potential successors, such as Rasul Guliyev and Ilham Aliyev— both representing the current Nakhichevan/*Yeraz* elite—are already on good terms with the United States, others have begun to seek patrons elsewhere. Since 1998, Yeni Musavat's Chairman Isa Gambar, has taken a more sympathetic turn towards Iran, away from his and his party's traditional western orientation.

The U.S. government must also continue its efforts to unify policy towards Azerbaijan. The continued enforcement of Section 907 hampers efforts to present a coherent, united policy front to the Azerbaijani government. Repealing Section 907 would immeasurably strengthen the Azerbaijani government's faith in American goodwill and would *not* represent a tilt toward favoritism of Azerbaijan. Rather, it would simply place Azerbaijan on an equal legal and diplomatic footing with all other post-Soviet countries, including the Lukashenko regime in Belarus that forced the American ambassador from his residence in 1998. Section 907 continues to deprive the United States of useful instruments (both tangible and symbolic) to influence events in Azerbaijan, and should be repealed. In neighboring Kazakhstan, in contrast, the provision of direct technical assistance, through the auspices of USAID and other agencies, has aided the government in the restructuring of the legal and administrative framework of its petroleum industry. Repeated in Azerbaijan, such a move would allow a deeper and broader engagement between American and Azerbaijani interests—both official and commercial—beginning in the country's oil patch. Ideally, this would allow American foreign policy to "piggy-back" on the inroads made by U.S. oil companies in Azerbaijan.

Finally, the United States should push more actively for a settlement of the Nagorno-Karabakh conflict. Such a move would likely yield dividends not only in fostering constructive relations with Azerbaijan, but also within the wider context of crafting a regional strategy for the Caucasus as well. As noted, the importance of Nagorno-Karabakh for Azerbaijan cannot be overstated. Both within the government and on the street, at virtually all levels of society, the lack of a settlement of the Karabakh conflict remains an open sore for most Azerbaijanis. Although there exist clear limits to what U.S. activism can accomplish on this front, any constructive effort towards crafting an acceptable settlement is likely to generate good will in Baku, regardless of who is in power. Whether under or outside of the auspices of the current OSCE-led negotiations, an assertive, non-deferential American presence can at least convince Baku, Stepanakert, and Erevan that a lasting settlement in Karabakh is as much a national priority for Washington as protecting Caspian energy reserves. Indeed, with the Karabakh front less than 20 miles away from the western oil pipeline hub at Yevlakh, such is as it should be.

NOTES

1. See chapter 3, "Crude Calculations: OPEC Lessons for the Caspian Region," by Terry Lynn Karl.

2. The timing of this decision—at the height of Russian pressure on the Azerbaijani government not to sign what would be called the "contract of the century" with the U.S.-led Azerbaijan International Operating Company (AIOC)—combined with the fact that Azerbaijan does not share a common border with Chechnya, casts considerable doubt on Russia's claim.

3. Officially, Azerbaijan's GDP grew at an incredible 8 percent in 1998. However, interviews with state statistics committee members and others, as well as numerous site visits, makes doubtful the accuracy of this number and the methodology used in generating it. Visits to several Absheron-area oil equipment factories, for example, yielded the following observation: officially reported workforce and production-output figures, on average, tended to be between four and ten times more than what was actually found on the factory floor.

4. The oil equipment sector in Azerbaijan, for example, suffered a calamitous drop in 1992–99 as orders from West Siberian oil producers disappeared.

5. With obsolete technologies, low productivity, and rising input costs, the non-oil industry has suffered in Azerbaijan. Of the metallurgy, machinery, chemical, and construction industries, none exceed 5 to 10 percent of their 1990 output levels. Lale Wiesner, ed., *Azerbaijan Economic Trends, First Quarter 1998* (Brussels: European Commission NIS/Tacis services, 1998). The strength of the Azerbaijani currency, the *manat,* has amplified these problems by making Azerbaijani exports more expensive to potential foreign buyers.

6. Iran's export-oriented religious beliefs and mistreatment of its ethnic Azeri minority (approximately 14 million ethnic Azeris live in northwestern Iran), coupled with the Azerbaijani government's westward orientation and occasional calls for reunification with "southern Azerbaijan," have sullied relations between the two countries. Especially damning, from Baku's perspective, is the perception of Russian-Iranian cooperation against Azerbaijani interests.

7. In 1995, OPON (*Otriad Politicheskii Osobykh Naznachenii*) special police forces under the command of then–first deputy foreign minister Rovshan Javadov and his brother Mahir staged a mutiny and attempted coup d'état. Immediately thereafter, when word emerged that the mutineers enjoyed material and moral support from elements of the Turkish government, the Turkish ambassador hastily departed Baku. Rovshan Javadov was killed during the fighting, and Mahir fled into exile in Austria. In early 1999, Mahir reappeared in Tehran to announce the formation of an armed movement with the express aim of overthrowing the government in Baku.

8. Section 907 was passed with the heavy support of the Armenian diaspora lobby in the United States, and was intended to pressure Azerbaijan into lifting its "blockade" of Armenia. Given the de facto state of war between Azerbaijan and Armenia at the time, it is difficult to envision exactly how this might have been accomplished.

9. In an interview with the *Washington Post*, SOCAR Vice President (President Aliyev's son), Ilham Aliyev, was asked whether oil might serve as a political weapon, to which he replied: "Yes. We have no strong diaspora, lobby, or sources for obtaining funds. We rely, therefore, on ourselves and . . . oil." (Reprinted in "Washington Post Interview with Ilham Aliyev, First Vice President of SOCAR and Member of the National Assembly," *Bakinskii Rabochii*, May 1, 1998, 3).

10. Independent decision making structures for Azerbaijan's energy policies only fully coalesced with the formation of the state-owned oil company, SOCAR, in September 1992.

11. From 1991 to 1993, for example, Azerbaijan was led, for varying lengths of time, by five presidents: Ayaz Mutalibov, Yagub Mamedov, Isa Gambar, Abulfaz Elchibey, and Heydar Aliyev.

12. It bears noting that this perception was shared by both the Aliyev and Elchibey governments.

13. The contract was originally scheduled to be signed in 1993 with the Elchibey government.

14. The Azerbaijani government and AIOC, after subtracting costs, expect to realize approximately $2.5 billion per year at peak output. Bonus payments for contract signings, meanwhile, have ranged as high as $300 million (AIOC paid a total of $300 million in bonus payments, of which roughly $70 million was credited for money paid to the previous Elchibey government, $60 million was deferred, and $90 million was used to repay Pennzoil for its gas compression project).

15. The Russian government has now admitted that it provided approximately $1 billion in armaments to the Armenian government between 1994 and 1997, and it continues to provide advanced weaponry, including MiG-29 aircraft and S-300 surface-to-air missile systems. The presence of Armenian regular army troops and equipment within Karabakh and other occupied parts of Azerbaijan, meanwhile, has been documented by foreign organizations and observers on numerous occasions, the denials of the Armenian and Karabakh authorities notwithstanding. See, for example, Human

Rights Watch/Helsinki, *Azerbaijan: Seven Years of Conflict in Nagorno-Karabakh* (Los Angeles, Calif.: Human Rights Watch, 1994).

16. Although the exact number of Azerbaijani refugees and internally displaced persons from Armenia, Nagorno-Karabakh, and occupied areas of Azerbaijan proper remains disputed, it is generally held to be between 800,000 and one million people. The organization Human Rights Watch, for example, refers to the forcible displacement of 450,000–500,000 Azerbaijanis from Armenian offensives conducted in 1993 alone. See Human Rights Watch/Helsinki, *Azerbaijan: Seven Years of Conflict in Nagorno-Karabakh*, viii.

17. Recruitment of ethnic Lezgins into the Azerbaijani army has been especially difficult, and has prompted a backlash of occasionally violent resistance by the Lezgin population in northeastern Azerbaijan. Meanwhile, relations between the central government and the Talish ethnic minority in the south have been strained since a short-lived attempt to establish an independent Talish state was crushed in 1993.

18. A term accorded by international observers to elections that are "neither free nor fair."

19. The eight opposition seats after the 1995 elections broke down as follows: three seats apiece for the National Independence Party of Etibar Mamedov and the Popular Front Party of former President Abulfaz Elchibey, and two for Musavat. In 1998, however, supporters of Rasul Guliyev within the ruling New Azerbaijan Party formed the Democratic Party and now claim six members of Parliament in their ranks.

20. Husseinov, a wealthy wool merchant turned military commander, led the military insurrection that forced the Elchibey government to collapse in June 1993. He served as Aliyev's first prime minister, but was later exposed as a Russian operative.

21. Author interviews with Azerbaijani journalists, bureaucrats, and government and opposition officials. Baku, Azerbaijan, August 1996 and February–July 1998.

22. On November 8, 1998, during an opposition rally protesting the alleged falsification of the October 11 presidential elections, three leading opposition figures—Abulfaz Elchibey, Lala Shovkat Hajieva, and Ilyas Ismailov—were attacked by a group of young men dressed in sports warmup outfits, seemingly with the assent of the nearby police. Later eyewitness reports linked members of this group to the "Amay" company, run by Jalal Aliyev, brother of Heydar Aliyev. If true, this incident brings into graphic relief the ties that bind economic, political, and kinship interests at the apex of Azerbaijan's ruling elite. Author's telephone conversations with local eyewitnesses and journalists in Baku, November 1998.

23. Minister of Defense Safar Abiev, for example, has reportedly been involved in a number of illegal business deals involving the sale of state property for personal profit. See "Defense minister to be fired by end of year," *Azadlyg*. October 23, 1999, pp. 1, 6.

24. Despite the fact that the *Milli Majlis* is officially required to review and approve all oil contracts, opposition deputies have complained that negotiations for most contracts are conducted "behind closed doors," and that the destination for bonus payments to the government is "totally unknown" to the Parliament.

25. Wages in the state sector in Azerbaijan averaged 116,381 manat (approximately $30) for the first quarter of 1998—approximately one-third the figure for the financial sector. Wiesner, ed., *Azerbaijan Economic Trends*.

26. This amount was to be recovered through a portion of subsequent AIOC oil sales.

27. Clearly, Baku's interests in this case were to preserve the volume of "profit oil," since that is where the bulk of the government's revenues are generated.

28. As evidenced by the political dominance of the *Yeraz* and Nakhichevan elite.

29. The Lezgin separatist movement Savdal, after a four-year lull, resumed active operations in early 1999, renewing calls for a Lezgin state straddling the border of Dagestan and Azerbaijan.

30. Author interview with Abulfaz Elchibey, Baku, March 28, 1998.

31. See, for example, "New Vatican," in the influential Baku weekly, *Ayna/Zerkalo,* May 1999.

32. The one opposition leader who has proven able to exploit this issue effectively has been Etibar Mamedov, leader of the Azerbaijan Party of National Independence. He is also the most prominent opposition figure *not* to have served in the AzPF government under President Elchibey.

33. It is widely believed that President Ter-Petrossian's ouster was predicated on his acceptance of an OSCE-brokered peace deal wherein six occupied regions of Azerbaijan would be returned pending negotiations on the final status of Nagorno-Karabakh. See, for example, Stephen Kinzer, "Ethnic Conflict in Caucasus Shows its First Glimmer of Hope." *New York Times,* September 14, 1998, p. 6.

34. Quoted in Terry Lynn Karl, *The Paradox of Plenty* (Berkeley: University of California Press: 1997), 4. See also her chapter 3 in this volume, "Crude Calculations: OPEC Lessons for the Caspian Region."

35. This was the last full year prior to the crash of oil prices on the world market in 1998, which, as of the first six months of 1998, had lowered oil-related revenues' share of total budget revenues to 32 percent. Source: Azerbaijan State Statistics Committee. Oil-related revenues in this case include oil bonuses, royalties, value-added tax, income tax, profit tax, excise, "other taxes," but not customs-collected taxes.

36. This number, however, is sensitive to oil-price fluctuations on the international market.

37. The creation of a Ministry of Fuel and Energy is unlikely to significantly undermine the nearly seamless ties between the president's office and SOCAR. Rather than serving as an independent generator of strategic energy policy, the MFE is more likely to assume regulatory functions by setting tariff rates, standardizing equipment, and carrying out other reforms. The importance of personal and kinship relations, however, means that much will ultimately depend on where individuals such as Ilham Aliyev end up.

38. While the official monthly salary for government administration employees was, as of mid-1998, 116,381 manat (approximately $30.61, according to the Azerbaijan State Statistics Committee), Azerbaijani specialists indicate that the price of a minimum subsistence "food basket" of goods at the time cost 224,650 manat per month. To provide a working person with a diet of 2,600 calories per day at the cheapest bazaar in Baku, moreover, would cost at least 400,000 manat per month. K. Ali and T. Arif, "My zhivem v poltora raza luchshe, chem nas uveriaiut." *Ayna/Zerkalo,* January 25, 1999, pp. 1–4.

39. In one humorous, yet far from unique, example, a high-ranking officer in the Baku police used his position to extort payments from foreign residents for the service of having his men "protect" their automobiles at night on the street. At the same time, he arranged to have the air removed nightly from the tires of said automobiles in

an effort to persuade their owners to move their cars into the newly opened, more expensive, parking lot across the street, owned and operated by the officer.

40. At several of President Heydar Aliyev's campaign rallies prior to the October 1998 elections, a large percentage of the "spontaneous" crowds in attendance were various civil servants—teachers, metro workers, bureaucrats, etc.—who were, according to them, bussed in by the authorities. Author interview with Azerbaijani journalists, October 29–30, 1998.

41. Specifically, Dutch Disease theories predict that the oil "wealth" that results from uneven investment into the oil sector will be appear as higher wages paid in the energy sector (thus attracting workers from other sectors) and will be spent on non-traded goods, thus leading to a real exchange rate appreciation that reduces the competitiveness of non-energy sectors. See W. Max Corden, "Booming Sector and Dutch Disease Economics: A Survey," *Working Paper 079* (Canberra: Australian National University, 1982).

42. Although the government has proven capable of slashing inflation and expenditures, this does not guarantee success in rationalizing future fiscal policy.

43. Figures based on data from first eleven months of 1998. Wiesner, ed., *Azerbaijan Economic Trends.*

44. As humorously referred to by U.S. Secretary of Energy Bill Richardson, "Building Caspian Solutions" conference, Washington, D.C., December 5, 1998.

45. "Happiness is multiple pipelines" was chosen by the U.S. government as the unofficial motto of its campaign to promote a multiple-pipeline, East-West energy transportation corridor, including trans-Caspian gas and oil pipelines and a large-diameter oil pipeline from Baku to the Turkish Mediterranean port at Ceyhan. The phrase, in addition to punctuating virtually all speeches on the subject by top U.S. officials, has found its way onto a widely-distributed bumper sticker as well.

46. British Petroleum, for example, has consistently favored the northern (Russian) route for AIOC exports from Azerbaijan.

47. The U.S. government continues to balk at subsidizing the construction of any east-west pipeline projects, although it has authorized the Export-Import Bank and Overseas Private Investment Corporation to provide financing.

48. Additionally, NATO does not maintain any permanent bases in non-member countries.

49. As evidence of the tenuous nature of Aliyev's health, he later underwent heart bypass surgery in Cleveland on April 29, 1999, again setting off a near political crisis in Azerbaijan.

5

Kazakhstan: The Long-Term Costs of Short-Term Gains

Pauline Jones Luong

With the full confidence of energy abundance behind him, Nursultan Nazarbaev, president of the Republic of Kazakhstan, wrote in the fall of 1997: "I am convinced that by 2030 Kazakhstan will become a Central Asian snow leopard and will serve as an example to other developing states. . . ."[1] In the years since independence, government officials have expressly relied upon the promised wealth of Kazakhstan's immense oil and gas reserves as the solution to its most acute social and economic problems, and as the key to its future development.[2] This pervasive attitude is not surprising, considering the emphasis that foreign governments and investors, international lending organizations, and development agencies alike have placed on Kazakhstan's energy sector. The United States government in particular has concentrated the bulk of its efforts in Kazakhstan on encouraging a legal, political, and economic environment conducive to oil and gas development.[3]

Nor is it surprising, considering the experiences of other countries endowed with substantial energy reserves, that the Kazakhstani government has pursued policies based on an overly optimistic estimation of its future income from the energy sector.[4] As in other petroleum-rich states, government officials' exaggerated calculations of impending riches have led Kazakhstan's leaders to encourage popular expectations of imminent wealth and to engage in overspending and borrowing. Meanwhile, they have turned their attention away from the plight of other key economic sectors, such as manufacturing and agriculture, and have failed to develop a reliable tax collection system.

At the same time, the Republic of Kazakhstan's strategy toward the development of its energy sector departs significantly from that of other states similarly well-endowed with natural resources. Whereas most such states choose to jealously guard these resources through full state ownership and control,

Kazakhstan rapidly privatized the bulk of its energy sector and invited an unprecedented level of direct international involvement in the development, production, and export of its energy reserves.[5] These policies have also generated some serious social, political, and economic consequences, including a strong reliance on foreign companies for improving socioeconomic conditions at the local and regional level, intense elite competition over the additional resources these foreign companies can provide, corruption at all levels of government, and an increasing trend toward the centralization and concentration of power in the presidential *apparat*.[6]

Why would the Kazakhstani government adopt such counterproductive policies? In short, because they offer several real and perceived short-term gains from the perspective of government officials.[7] The pervasive optimism that oil and gas reserves would bring unbridled prosperity to Kazakhstan has helped to diffuse widespread popular discontent and to attract the attention of foreign investors and governments. Moreover, it has helped to secure aid from international development agencies such as the U.S. Agency for International Development and the European Union's TACIS program, which provides financing for projects that promote market economies and democratic societies in the former Soviet Union and Mongolia, and lending institutions such as the International Monetary Fund, World Bank, and the European Bank for Reconstruction and Development. In addition, large-scale privatization satisfied the government's pressing need for revenue in the wake of the Soviet collapse. Revenues from privatization helped fill the budgetary gap once Moscow stopped transferring resources to Kazakhstan, provided much needed capital and socioeconomic investment at the local and regional levels, and enabled the delay of democratic reforms. Due to these policies, Kazakhstan was also able to achieve some positive economic results from 1996 to 1998, including macroeconomic stabilization, a significant decline in inflation, and a modest growth in GDP.[8]

Unfortunately, these short-term gains conceal both the short- and long-term costs for the country as a whole. Some of these costs are already materializing in the form of mounting public protests and a shrinking economy. The full extent of the harm inflicted by such short-sighted policies, however, will only become apparent over the next decade. These negative outcomes could include the intensification of poverty, unemployment, popular discontent, elite competition, and general economic decline. In addition, if current trends continue, Kazakhstan will emerge as a quasi-state—that is, one with international legitimacy but without the domestic capacity to generate sufficient revenue, address basic social problems, and promote even minimum levels of economic growth.[9]

In sum, Kazakhstan's approach toward its energy sector is aimed at promoting political acquiescence and providing social and economic relief in the short-term, yet it is actually increasing the likelihood for political instability and socioeconomic decay over the long term. (See table 5.1.) This is the crux of the argument substantiated in the pages that follow.

Table 5.1: Short-Term and Long-Term Effects of Kazakhstan's Energy Wealth

	Short-Term Effects	Long-Term Effects
Over-Optimistic Expectations	1. increases popular expectations 2. promotes over-spending & borrowing; increases discretionary spending 3. decreases attention to the development of other economic sectors; increases unemployment 4. decreases emphasis on tax collection	1. will exacerbate popular discontent & further erode support for reform 2. will impede economic growth & undermine state capacity 3. will promote Dutch Disease & increase poverty 4. will impede state capacity to generate revenue & hinder democratization
Form & Pace of Privatization	1. foreign companies take responsibility & blame for local socio-economic conditions 2. quick sell undervalues natural resources; large royalties provide quick revenue, increase intra-elite competition 3. fosters corruption at the central and regional levels, popular distrust of government officials 4. encourages trend toward centralization & concentration of power	1. will increase distance between central government & local population, promote state incapacity, & exacerbate regional socio-economic differentiation 2. as revenues decrease over time, elite competition will intensify 3. will impede democratization & hinder economic growth 4. increased center-regional tensions, impede democratization, & hinder economic growth
Combined	1. decreases state commitment to direct social spending, increases poverty & health risks 2. creates repeated "cycles" of protest, crackdown, and co-optation	1. will exacerbate growing regional economic differentiation, greater poverty & health risks 2. may lead to the radicalization & regionalization of the opposition

THE EXPECTATION OF OIL AND GAS RENTS:
SHORT-TERM EFFECTS

Immediately after Kazakhstan's independence, government officials embraced the international community's faith in its immense energy wealth. The reserves in the Tengiz field were initially estimated to hold up to ten billion barrels of oil, while the other two large onshore oil fields in western Kazakhstan (Uzen and Karachaganak) were estimated to hold approximately one-third of this amount each. Domestic politicians and western investors alike quickly accepted these figures, predicted output and income accordingly, and minimized problems associated with improving production and securing viable export routes. Chevron, which established the Tengizchevroil joint venture in April 1993 as the first of its kind in Kazakhstan, expected that it could eventually produce over 700,000 barrels per day for export.[10] Yet, although production has indeed increased every year since then, as of 1998 it had still failed to reach this level and remained constrained by limited export access.[11] Similarly, the $150 billion that President Nazarbaev claimed his country would earn in just a few years from Tengiz exports has not yet materialized.[12]

The widespread enthusiasm for Kazakhstan's onshore reserves was minuscule compared to the excitement generated by the potential of its offshore reserves in the Caspian Sea. In June 1996, based on a series of seismic tests conducted by the Caspian Sea Consortium, Kazakhstan presented official estimates of 73 billion barrels of crude oil and 2 trillion cubic meters of natural gas.[13] This meant that Kazakhstan's sector of the Caspian held reserves "ten times bigger than those of its onshore Tengiz oilfield and [much larger than] Russia's entire oil reserves of 6.7 billion tons [48.9 billion barrels]."[14] Once again, political and structural difficulties were minimized in the face of such awesome potential. Kazakhstan's oil and gas minister Nurlan Balgimbaev himself declared at the end of 1996 that drilling would begin the following year, and that the revenue would flow to the central budget shortly thereafter.[15] Likewise, Baltabek M. Kuandykov, president of Kazakhstan- caspishelf, the consortium of Kazakhstani and foreign companies that undertook a survey of Kazakhstan's oil deposits in the north and east part of the Caspian Sea, announced that the government expected "the Caspian oil quantity to be enormous compared with the Tengiz reserves and by the year 2015 to supply "60 million tons [438 million barrels] of oil per year."[16] More recent estimates of Kazakhstan's Caspian reserves, however, are as low as 7.7 billion barrels for the entire region.[17]

Not even the continued decline in oil prices from mid-1997 to mid-1999 diminished the Kazakhstani government's optimism or encouraged officials to reconsider its reliance on oil production and export. In his new position as president of Kazakhoil (the state oil and gas holding company formed in March 1997), Kuandykov insisted in March 1998 that while "at present his country's oil exports were unaffected by the fall in prices . . . , further price falls could lead to a reduction in Kazak exports in the near future."[18] Over a year later, however, these reductions still failed to materialize. As of mid-1998, all Kazakhstani oil companies continued to increase production, and Chevron and its partners announced plans to boost production at Tengiz.[19] Over a year after oil prices began to drop, President Nazarbaev continued to boast of the attractiveness of international oil and gas development projects in his country and to express confidence in the country's long-term economic growth prospects.[20] In early 1999, Nurlan Balgimbaev, who had since been promoted to the position of prime minister, declared that by the following year the Republic of Kazakhstan intended to produce 292 million barrels of oil annually, 204 of which would be reserved for export, and to more than triple this amount by the year 2015.[21] This is a more optimistic estimate of Kazakhstan's oil export capacity than that of most international oil experts.[22]

While the fervent optimism of the Kazakhstani government regarding the potential of its energy reserves is well known by now, what is much less widely recognized is that this perspective has had several negative social, political, and economic effects. In this regard, Kazakhstan has experienced outcomes that are common across petroleum-rich states both within and outside of the Caspian Sea region, as the chapters in this volume by David Hoffman and Terry Lynn Karl make clear.

First of all, grandiose visions of oil wealth have raised popular expectations that oil and gas reserves can serve as the panacea for all that ails Kazakhstan. Government officials, most notably the president himself, have not hesitated to share their overly optimistic expectations with the public. In February 1997, for example, President Nazarbaev declared at an official meeting of national political leaders and other elites that the Republic of Kazakhstan would receive some $76 billion from the Tengiz oil field project, and that Kazakhstan would take in 76 percent of all revenues from the Tengizchevroil joint venture. He also predicted that these funds would begin to appear in the very near future, along with even greater amounts pouring in from contracts to develop oil reserves in Kazakhstan's section of the Caspian Sea: "By the start of the century, this money will not only arrive, but also significantly increase in volume. . . . Kazak heavy industry will rise up and mineral fields will come into development, including 180 gold fields. . . ."[23]

Similarly, in connection with his official visit to the United States in the fall of 1997, Nazarbaev publicly acknowledged his expectations that the latest oil agreements signed in Washington envisaged a long-term investment of $28 billion, from which Kazakhstan would receive up to $550 billion in oil revenues over the next forty years.[24] He later declared during a visit to Pavlodar refinery as part of his presidential re-election campaign in November 1998 that the country would be a major oil and gas exporter in the next century.[25]

Second, the overly optimistic view of future oil revenues has promoted a strong tendency toward borrowing and overspending, particularly for discretionary purposes. By 1997, Kazakhstan was already the third largest borrower among the former Soviet republics (after Russia and Ukraine), having accepted 13 loans from the World Bank totaling roughly $960 million since independence.[26] A large portion of this was directed initially toward macroeconomic stabilization and then redirected toward the rehabilitation of oil fields and the reorganization of state-owned enterprises to attract foreign investment. Meanwhile, more direct income from the export of oil created an extra-budgetary source of discretionary funds. The creation of Kazakhoil in particular made this possible, since it provided a legal means for the government to set aside a portion of the company's account acquired through oil exports to satisfy its own aims. Financing for the construction of new buildings and infrastructure in the new capital, Astana, is a case in point. Kazakhoil is estimated to have spent $25–30 million on improvements in Astana.[27] The president of Kazakhoil himself admitted that "a major portion of the oil dollars are being invested in the construction of the new capital."[28] Over the next several years, the transfer of the capital to Astana is likely to "cost at least $500 million and—with a shortage of power and drinking water—will eventually run up a still larger tab."[29] The government's expectation is that these funds will come from oil dollars.

The third negative consequence of the Kazakhstani government's reliance on the energy sector is that it has shifted official attention away from the development of other sectors, most notably manufacturing and agriculture. The investment priority of the government since independence has clearly been in revitalizing its oil and gas fields and building pipelines. Between 1990 and 1995, the oil industry alone accounted for 40 percent of all sectoral investment; the energy sector (including electric power generation and coal) made up 60 percent of total sectoral investment.[30] According to an IMF report, by 1998, "[c]apital investments in the national economy [had] shrunk by 89 percent compared to 1991, nine out of every ten industrial enterprises [had] stopped work, and livestock numbers [were] half of what they used to be in 1991," yet investment in the energy sector continued to grow by comparison with other sectors.[31]

This, in turn, has led to a significant rise in unemployment, the out-migration of skilled labor, and a growing sense among the population that the government is simply ignoring its plight. Between 1991 and 1996, for example, "the number of wage employees [in the industrial sector] fell by 36 percent, . . . over three-fourths of the decline in wage employment in the economy as a whole."[32] Unofficial unemployment is likely to be even higher. Some local experts estimate that every fourth Kazakhstani citizen of working age has no job, and that this particularly affects those under age 30 and women, who make up 66 percent of the total unemployed population in Kazakhstan.[33] As one regional analyst remarked, the government's prioritization of its extractive industries has not only contributed to the collapse of food, petrochemical, machine-building, and other sectors, but also has left many skilled workers and their families without jobs or the prospects for finding one.[34] The direct result of this, he argues, is mass out-migration by skilled laborers in search of employment opportunities elsewhere.[35]

Yet perhaps more importantly, the failure of the government to address the demise of the industrial sector has contributed to a growing perception that government leaders have lost their sense of reality.[36] According to the prominent Kazakhstani human rights activist, Evgenii Zhovtis: "If 2–3 years ago the authorities tried to react to the problems of the industrial enterprises . . . , today the plight of whole cities . . . is of little interest to them. The authorities wait for when the situation will remedy itself."[37]

Finally, the expectation that the country can rely primarily on oil and gas revenue has diminished the government's efforts to develop a reliable system of tax collection. When state revenues declined to 20 percent of GDP by 1995, and continued to drop another 6 percentage points in 1996, government officials still believed that to solve deficit problems it would be easier to wait for oil revenue to start flowing than to improve tax collection or economic growth.[38] Although the government acknowledged its failure to collect taxes effectively during 1997, it has done little to correct this problem aside from creating a new State Revenues Ministry in October responsible for fiscal policy, tax regulation, and customs.[39] Tax revenues as a percentage of GDP continued to decline in 1998.[40]

THE REALITY OF OIL AND GAS RENTS: LONG-TERM CONSEQUENCES

By most western estimates, the actual income generated from Kazakhstan's energy wealth will be much smaller than expected by government leaders due to several factors including: (1) the high costs of transporting oil from a landlocked state to world markets; (2) the quality of the oil reserves themselves, some of which are high in sulfur; and (3) the overall downward trend in world oil prices since the 1980s.[41] Some western experts predict that the most likely scenario for Kazakhstan's projected oil rents, which estimates the world price of oil at $18 per barrel and Kazakhstan's output capacity at 1.5 million b/d, would produce only a modest rent stream, or approximately 1.8 percent of the country's GDP. This is much lower than the windfalls that other petroleum-rich countries, such as Venezuela, Indonesia, and Nigeria, experienced in the 1970s.[42] In addition, based on this scenario for projected rents, Kazakhstan's actual revenue from oil export would boost government income by only 10 percent (assuming the level of public expenditures stabilizes at 16 percent of GDP).[43] Thus, despite the upswing in world oil prices after OPEC members pledged to reduce output in mid-March 1999, oil export alone is unlikely to produce an economic boom in Kazakhstan. This upward trend could also prove ephemeral either if OPEC is unable to restrain members from the temptation to "cheat" or if production continues to increase among non-OPEC countries due to increased yields from existing reserves through technological advances.[44]

The failure of the influx of oil and gas revenues to benefit the vast majority of the population has already contributed to growing popular discontent about the state of Kazakhstan's economy, and consequently to the erosion of support for market reform. Recent polls conducted by the Giller Institute in Kazakhstan between 1995 and 1997 indicate that only about 30 percent of the population supports Nazarbaev's chosen pace and type of economic reform. Polls conducted by the International Foundation for Electoral Systems in 1996 found that almost half of the adult population preferred to abandon the present course of reforms and return to a state-controlled economy.[45] And, while in 1995 the majority of respondents shared an optimistic view that economic conditions would improve greatly by the year 2000, in similar polls taken approximately two years later less than one-third of respondents believed Kazakhstan would be a "well-off" country at some point in the future, and just 1.9 percent felt the economic situation in Kazakhstan had improved.[46]

Moreover, by my own estimates, the number of recorded public demonstrations and strikes has increased approximately 17 times since mid-1995, and the unrecorded amount is likely to be even higher. These protests have occurred not only to call attention to Kazakhstan's deteriorating economic conditions but also to demand some relief. (See table 5.3.) This situation contrasts sharply with the virtual absence of social unrest that David Hoffman and Nancy Lubin describe in their chapters in this volume on Azerbaijan and Turkmenistan, respectively. It also points to some important differences in the economic and social make-up of Kazakhstan vis-à-vis these other Caspian littoral states. In particular, because Kazakhstan inherited a much larger industrial sector from the Soviet Union than the other Central Asian republics, it also has a much stronger coterie of labor unions. These workers' organizations have managed to survive, and indeed thrive, despite Kazakhstan's reversion to authoritarianism since 1995 and the general weakness of its civil society.[47] Over the long run, the reality of unrealized rents will certainly exacerbate this growing popular discontent and further erode support for market reform if the Kazakhstani government continues to pursue current economic policies.

The central government's tendency to borrow against future oil and gas revenue and to engage in over-spending will undoubtedly impede Kazakhstan's future economic and political development. As economist Richard Auty notes, "even modest oil revenues can severely damage economic prospects if governments are overly optimistic (as is currently the case in Kazakhstan) and, as a result, unwisely use the revenues as collateral to boost external debt."[48] In 1996, Kazakhstan was already spending nearly 10 percent of its budget revenues on servicing its external debt, and as a direct result dramatically cut all social programs.[49] Thus, high levels of foreign debt are likely not only to hinder economic growth but also to impede the state's capacity to perform basic functions, such as providing essential public goods and social services.

The official shift in focus toward the energy sector and away from other sectors, combined with the government's failure to bolster tax collection, will have other serious long-term negative economic and political consequences. Although unemployment remained relatively low in 1997, it has been on the rise since 1998 in the manufacturing sector and even began to spread into energy-sector enterprises as the force of falling oil prices squeezed profit margins.[50] The effects of rising unemployment are already being felt in declining tax revenues, rising poverty levels, lack of access to health care, and increasing out-migration.[51] Over the long term, not only are these trends likely to continue, but it is doubtful that the budgetary outlays required to address them will be adequate. For example, since unemployment is predominant among the young and poorly educated, and out-migration is occurring among middle-aged skilled laborers, this problem cannot be remedied by revitalizing industry alone.[52] It also will require massive government investments in vocational training and technical education.

Yet without adequate tax collection institutions, the government will continue to have difficulty generating revenues with which to make these investments. According to Murat Asipov, a local journalist, "poor revenue performance has [already] contributed to disproportionate declines in the social safety net, public investment, and the funding of public services."[53] As of 1998, "public expenditures [were] running at about *a third to a half* of pre-independence levels in real terms."[54] Ironically, the modest amount of rents that it is estimated Kazakhstan will earn, even with a slight upward trend in world oil prices, is far less than would be achieved through improved tax collection.[55] Taxation is also germane to the future of democracy in Kazakhstan. Without a viable tax system, it is unlikely that Kazakhstani citizens will ever develop the direct and reciprocal relationship with their state that has long been recognized, at least implicitly, as the basis for democratic governance.[56] The very purpose of democracy is to devise domestic institutions that effectively channel societal demands to the country's leadership.[57] These leaders are compelled to be responsive to societal demands, not only because they depend upon societal support to retain public office, but also because their position and ability to govern depends upon the payment and collection of taxes. Taxation has also historically served as a basis for demanding popular enfranchisement. One only need recall the cry of the American revolutionaries—"no taxation without representation"—to grasp the significance of this link between society's economic contribution to the state and its political leverage.[58] By relying on foreign rather than domestic sources of revenue, Kazakhstan's leaders can afford to delay democratization and neglect societal interests in pursuit of their own. The experience of other resource-rich states is also instructive here, since the majority of them have followed development trajectories similar to the one on which Kazakhstan appears to have embarked.[59]

FULL SCALE AND FAST-PACED PRIVATIZATION: SHORT-TERM EFFECTS

In short, the Republic of Kazakhstan's energy sector development strategy has consisted of attracting as much foreign investment and as many foreign investors as quickly as possible. This was an expressed goal of former Prime Minister Akezhan Kazhegeldin, the steward of the country's privatization program, who was able to secure $1.2 billion in direct investment in the first year alone.[60] Under his leadership, the Kazakhstani government sold off the bulk of its shares in oil and gas enterprises to a variety of foreign companies between July 1996 and July 1997. (See table 5.2 for details.) Thus, by 1998 Kazakhstan was the largest per capita recipient of foreign direct investment in the Commonwealth of Independent States, most of which is heavily concentrated in the energy sector.[61]

The central government clearly viewed the rapid sale of oil and gas enterprises as an opportune way to raise desperately needed funds and to expand and modernize rapidly the country's production, refining, and transportation capabilities.[62] Unfortunately, the government's expectations for the amount of revenue that would accrue via privatization were much higher than the actual amount received. Many government officials were dissatisfied with the prices obtained by Prime Minister Kazhegeldin in the first few rounds of privatization. In fact, this was a major reason for his forced resignation in October 1997. Tenders handled after his departure by the newly appointed Prime Minister (and former Oil and Gas Minister) Nurlan Balgimbaev thus aimed for much higher numbers and for more money up front.[63] In a recent sell-off of state shares in the Offshore Kazakhstan International Operating Company (OKIOC) to Phillips Petroleum of the United States and Inpex of Japan, for example, Balgimbaev declared that Kazakhstan would receive "immediately . . . around $500 million" and eventually would "reap $700 billion in revenues from its offshore oil and gas fields . . . , 90 percent [of which would] go to Kazakhstan in royalties, taxes and bonuses [and] the remaining 10 percent . . . to the foreign companies."[64] Most foreign experts and oil and gas company representatives working on the ground, however, insist that a much more modest amount of revenue will accrue directly to the Kazakhstani government.[65]

The Kazakhstani government also had bloated expectations concerning the

Table 5.2: Energy Sector Privatization in Kazakhstan

Date	Foreign Company	Fields or Enterprise	% Shares	Official Price
July 1996	Vitol Munay	Shymkent Oil Refinery	85%	$230 million[a]
August 1996	Hurricane Hydrocarbons	Yuzhneftegas JSC	90%	$400 million[b]
August 1996	Tractebel	Almatyenergo	100%	$275–698 million[c]
March 1997	Triton-Vuko Energy Group	Karazhanbasmunai	94.5%	$90 million
May 1997	Medco Energy Corporation	Mangistaumunaigaz JSC	85%	N/A
June 1997	Chinese National Petroleum Co.	Aktyubinskmunaigaz JSC	60%	$4 billion
July 1997	Chinese National Petroleum Co.	Uzen Oil Fields	60%	$9.5 billion[d]
January 1998	Central Asia Petroleum	Mangistaumunaigaz JSC[e]	65%	$248 million

[a]This consists of $80 million in direct payments over three years and another $150 million in investments over five years to upgrade the plant. Vitol (a Dutch company) backed out of the deal in early 1997, and the contract came under the supervision of a former Vitol subsidiary, Kazvit.

[b]This includes a pledge to invest $280 million in the enterprise over six years.

[c]This lower figure includes $270 million pledged to upgrade several power stations, the electricity grid, and pipeline system; the higher figure is based on $340 million (prob) in expected payments and $358 million in direct investments, including a pipeline in southern Kazakhstan that will bypass Kyrgyzstan.

[d]Approximately half of this amount is pledged investment in a pipeline from these fields to eastern China.

[e]Mangistaumunaigaz JSC was re-sold to another Indonesian company after Kazakhstan established its own stock-holding company, Kazakhoil, in March 1997.

level of tax revenues it would eventually receive from foreign companies. For example, Kazakhstani government officials anticipated that they would earn an additional $2 billion in taxes and royalties beyond the initial purchase price over the next 20 years from the sale of a controlling stake in Yuzhneftegaz alone.[66] Yet ironically, most contracts with foreign companies included exemptions from tariffs and taxes.[67] In lieu of taxes, many foreign companies have agreed to make huge capital investments in technology and infrastructure, as well as to pay back wages, contribute to pension funds, and build roads, schools, apartments, and hospitals. Hurricane Hydrocarbons, the company that eventually won the tender for Yuzhneftegas, is a case in point. Written into its contract was a provision requiring it to assume all the social obligations and economic costs of the company and surrounding area, including $4 million a month in local salaries for workers (some of whom never actually existed).[68]

This common practice of foreign companies assuming social costs was not merely a means of temporarily relieving the government of its fiscal burdens in the regions, but also a way of overcoming the initial opposition to privatization from managers of oil and gas enterprises. Indeed, central authorities continually reassured both regional governors and factory managers that they would benefit directly from privatization—promising that the foreign companies would invest in the necessary technologies to increase production, pay social costs, and in many cases guarantee employment and the payment of back wages.[69]

The principal consequence of the privatization scheme, then, has been to place both the responsibility and blame for local socioeconomic conditions on foreign investors, rather than on government officials. The governor of Atyrau, for example, views Tengizchevroil as crucial to the region's economy, since it employs over 3,000 workers and provides more than a quarter of its tax revenues.[70] When Tengizchevroil attempted to reduce the number of its employees through a reorganization program that would transfer additional services to subcontractors, however, the regional government was quick to side with the local labor union leaders against this move.[71] Union leaders and workers alike have also come to hold the foreign owners responsible for their social and economic well-being, even though their respective enterprises were in dire straits well before privatization began.[72] Since the Belgian company Tractebel acquired the former state energy company Almatyenergo in August 1996, for example, it has repeatedly been held accountable for employee grievances and general resentment toward the failure of privatization to solve the company's deep-seated economic and technical problems.[73] Tractebel attempted several times to increase Almatyenergo's efficiency by raising rates and reducing its enormous labor costs, but instead had to appease local authorities and union leaders by agreeing to maintain its bloated workforce and pay wage arrears.[74]

The Kazakhstani government in fact deliberately has encouraged the tendency to blame foreign investors. When 1,000 workers who had not received their wages for over a year began a hunger strike in the fall of 1997, for example, central and local authorities claimed that the new owners (LUKoil-Utek joint stock company) were responsible for back wages. Thus, their proposed solution to this problem was to try to find a "more respectable" investor.[75] According to some foreign and domestic experts, a secondary consequence of Kazakhstan's policies, the speed of privatization, has led to the undervaluing of Kazakhstan's natural resources. When asked why foreign companies continue to invest in Kazakhstan despite problems with tax laws, corruption, and the lack of transparency in tender bids, foreign investors responded: "The prices are cheap. . . . In CIS countries and particularly Kazakhstan, the perception is that these assets are relatively undervalued. . . . There's now a real pressure from shareholders for the bigger returns from emerging markets and whoever gets in first usually gains."[76]

Undervalued assets together with the promise of large royalties and hidden "side-payments" (or bribes) from the sale of Kazakhstan's most valuable assets to foreign owners, has increased competition among governing elites. Competition for control over the privatization process itself has been particularly fierce between those who work within and those who work outside of the oil and gas sector.[77] The greatest opposition to privatization initially came from the so-called "oil barons" in western Kazakhstan who were unwilling to "voluntarily surrender" their long-held monopoly over the country's oil wells and gas pipelines. In fact, they were initially successful in delaying the oil and gas tender process.[78] In spring 1997, the central authorities eventually responded by dissolving the stronghold of the oil barons' authority—Kazakhstanmunaigaz, the state holding company that controlled the oil and gas sector—and continued with privatization as planned.[79] However, the government was subsequently forced to make some concessions to energy sector elites when promised revenue from oil and gas enterprise sales failed to materialize. Thus, in October 1997 Nazarbaev put one of the most prominent energy sector leaders, Nurlan Balgimbaev, in charge of slowing down the privatization process and reconsidering pending contracts.

The anticipation of additional revenue generated via privatization has also fostered corruption among both central and regional elites. Foreign companies must first negotiate contracts with central authorities—most notably, the president and prime minister—to whom they often offer "surplus funds" as a way of "sealing the deal."[80] They must then contend with the regional governors (or *akims*), with whom they engage in on-going negotiations. Not only do they need the *akims'* approval for the initial purchase, they also have to keep the governors involved in decision-making. Otherwise, *akims* can make it difficult for the foreign companies by causing delays in the project or sending over the tax inspectorate. Local officials are not opposed to accepting personal cash payments in exchange for helping the project run smoothly or for ignoring contract violations and environmental regulations.[81]

This corruption, in turn, has contributed to growing popular distrust of government officials. The popular Almaty-based newspaper *Karavan*, for example, accused the government of stealing money generated from privatization, including half a billion U.S. dollars paid by Mobil for shares in Tengizchevroil.[82] Opinion polls taken in 1997 and 1998 indicate that this is a widely shared view, and that as a result public trust in the Kazakhstani government is at an all-time low.[83]

Finally, since the spring of 1995, the drive to attract foreign investment in the energy sector has encouraged the Republic of Kazakhstan's trend toward the centralization and concentration of power. In March 1995, Nazarbaev dissolved a parliament that had become increasingly hesitant to endorse his economic reform agenda. This was followed by a referendum to extend Nazarbaev's term to the year 2000, decrees aimed at augmenting the power of the presidency over all economic and foreign policy matters, and other presidential actions. These events laid the groundwork for another wave of economic "reforms" in March 1997 which were intended to facilitate foreign investment, particularly for newcomers. The sum result of these reforms was to establish a "one-stop shopping center" for foreign investors in all areas of the economy, but particularly in the energy sector.[84] Similar to the relationship between SOCAR and western oil companies that David Hoffman describes in Azerbaijan, foreign investors have welcomed the centralization of political and economic decision making to expedite their business transactions.

THE PUBLIC COSTS OF PRIVATIZATION: LONG-TERM EFFECTS

Several long-term consequences stemming from privatization have also arisen due to the Republic of Kazakhstan's increased reliance on foreign companies for regional socioeconomic welfare. In the short-term, this strategy has undoubtedly alleviated some of the country's most pressing problems, especially in regions like Kyzl-Orda which were hit hardest by lost subsidies after the Soviet Union's demise. Yet over the next decade, Kazakhstan's privatization program is likely to increase the distance between central authorities and the local population, to exacerbate regional socioeconomic differentiation, and ultimately to promote state incapacity. There is already a growing sense in Kazakhstan that central leaders are detached from the realities of everyday life. According to a well-known local political commentator: "the gulf between [central] authorities and citizens has grown. . . . The authorities do not understand the situation adequately. . . . The leadership of the republic feels a euphoria because of achieved economic successes. Actually, the successes are not at all clear."[85]

If foreign companies continue to assume responsibility and bear the blame for Kazakhstan's social and economic condition, the central government will become increasingly removed from the daily problems that its citizens face. Perhaps this would not be so worrisome if it did not also coincide with the weakness of institutional mechanisms—such as fully contested elections and active political parties with a strong popular support base—for transforming societal demands into policy outcomes.[86] The presence of such factors would contribute to more responsive and effective governance in Kazakhstan; their absence will promote precisely the opposite result.

Moreover, what the central leadership considers its "economic successes" are felt differently across Kazakhstan's regions. There are already striking differences in the poverty rates between the southern, eastern, and the northern parts of the country: "The poverty rate in the south of the country (69 percent) is nearly twice the national average, while that in the north (9 percent) is many times lower. In terms of the number of poor, nearly two out of three poor people live in the south or east of the country."[87]

Similarly, levels of consumption vary across regions, which are highest in the west and lowest in the north.[88] Such stark differences in regional economic development are especially alarming because they are likely to exacerbate tensions stemming from Kazakhstan's overlapping ethnic and territorial divisions. In contrast to the other post-Soviet states discussed in this volume, Kazakhs have not constituted a majority in their titular republic-turned-independent-state since 1926. Since then, Kazakhs and Russians each have made up approximately 40 percent of the population.[89] Owing to both historical and political factors, over time Kazakhstan became geographically divided between Russians and Kazakhs: Russians became a majority in most of the northern and eastern parts of the republic, while Kazakhs remained numerically strong in the south and west.[90] These overlapping ethnic and territorial divisions have been an important source of social and political tensions in present-day Kazakhstan. Indeed, shortly before independence, Russian nationalist movements emerged in the northern and eastern regions to demand greater autonomy or outright secession, while Kazakh groups in the southern and western regions rallied around elevating the status of Kazakh to the official state language and appointing more Kazakhs to government positions. Several years later, Russian and Kazakh nationalist organizations continued to draw their primary support base from the northern and eastern oblasts or southern and western oblasts, respectively.[91]

As the growing levels of social unrest since 1996 seem to indicate, in some cases these so-called "economic successes" are not felt at all.[92] (See table 5.3.) Workers and union leaders, in particular, have become increasingly skeptical that privatization will bring the promised rewards. Some foreign companies have not paid back wages and maintained full employment, as the government promised.[93] Labor has thus begun to unite, not only to protest this situation but also to demand that the government actually remedy it.[94] In this sense, privatization has generated something positive, an organized collective action. Yet, to date such popular demands for redress have not produced a positive government response: as table 5.3 illustrates, thus far the government has primarily either ignored or repressed them. However, it seems unlikely that the government will be able to continue its current approach, and this deadlock constitutes yet another source of potential future conflict.

Even if foreign companies are willing and able to meet the costs of the socioeconomic burden that contracts with the Kazakhstani government have bestowed upon them, this will create more problems rather than provide a solution. Over time, the state itself will become unable to meet the basic needs and demands of its population because it has not developed the institutional capacity to do so. An increased reliance on foreign companies to perform what are considered state functions will thus undermine the state itself. State-society relations will not be mediated directly through domestic institutions, but rather indirectly via international actors and organizations. In short, Kazakhstan will become a state that essentially governs by proxy.[95]

Table 5.3: Public Demonstrations and Strikes in Kazakhstan, 1995–1998

Date & Place	# Of Participants	Cause	Government Response
July 18, 1995; Northern Kazakhstan	Approximately 1,300 miners	National government announced plans to close down a number of inefficient coal mines	National government ignored the demonstrations; Local government down-played them
January 13, 1996; northern Kazakhstan	Over 100,000 coal miners	Demanded the government pay several months' back wages	Refused to yield to "economic blackmail"
December 8, 1996	Approximately 3,000 people	Deteriorating economic conditions; Demanded economic reforms cease and the government resign	Issued warnings to group leaders that they would be prosecuted
Beginning January 21, 1997; South and North Kazakhstan	Not available	Demanded payment of wage arrears	None
February 6, 1997; Semipalatinsk	1,500 teachers	Demanded payment in full of 400 million tenge (about $5.3 million) in wage arrears	None
March 17, 1997; Almaty	Approximately 300 pensioners	Commemorated the 6th anniversary of the all-Union referendum on preserving the Soviet Union	None
March 23, 1997; Kentau to South Kazakhstan (march from Almaty)	Not available	Non-payment of wages for over 6 months	Local authorities stopped the marchers about 15 kilometers from Almaty; Central authorities warned that such protests were illegal and called participants criminals
March 24, 1997; Almaty	Approximately 300 people	Worsening living conditions	None

(Continued)

Date; Location	Number of participants	Demands	Government response
March 27, 1997; Kokshetau (northern Kazakhstan)	Local pensioners	Unpaid pensions	Warned by Kazakhstan's Procurator-General not to hold demonstration
May 30, 1997; Almaty	Estimated in the thousands	Demanded the resignation of parliament and the president, and the payment of back wages and pensions	None
June 17, 1997; Karaganda	Approximately 500 miners	Protested the country's pension system, which raised the eligible age from 60 to 63 for men and 55 to 58 for women	No direct response, but Nazarbaev promised less than a month later that all pensions would be paid
July 15–16, 1997; Akmola oblast and Kokchetau city	Not available	Protested wage arrears; demanded payment of wages and pension arrears, and a cut in utility costs	None
July 30, 1997; Kokchetau city	About 400 people	Called for an end to economic reforms, which they claimed were having little effect	None
July 31, 1997; Almaty	Not available	Protested pension arrears and rising housing costs	Parliament's lower house allegedly invited representatives to discuss their complaints
September 11, 1997; Western Kazakhstan	Approx. 100 employees of the Karazhambasmunai oil company	Demanded payment of late wages and protested company's plans to lay off employees	Local government sided with workers; demanded company pay back wages
September 30, 1997; Almaty	200 pensioners	Demanded increase in pensions and that the government pay more attention to their situation	None
October–November 1997; Kentau city (South Kazakhstan)	About 1,000 workers (from the Achisay Polymetal plant)	Protested declining living standards and demanded back wages	Made initial warnings, then physically stopped protestors from marching to Shymkent as

(Continued)

Table 5.3: Public Demonstrations and Strikes in Kazakhstan, 1995–1998 (*Continued*)

Date & Place	# Of Participants	Cause	Government Response
			planned; paid some wage arrears to workers not taking part in the protest march; once strikes ceased, wage arrears were paid to all workers with a 150 million tenge credit from the Kazakhstani government
October 8, 1997; Karaganda oblast	Not available	Protested the planned closure of several coal mines in the oblast	None
October 31, 1997; Almaty	Not available	Showed solidarity with the Achisay Polymetal plant workers in Kentau	None
November 5, 1997; Jambyl oblast	24 workers' representatives from the Qaratau phosphorus producing plant	Demanded payment of wage arrears	None
November 30, 1997; Almaty	Members of the opposition	Protested government's arbitrary rule	Warned leaders and then punished them with arrests and large fines

The attitude of elites toward privatization in general and foreign investors in particular will also generate some serious long-term political and economic consequences for Kazakhstan. First of all, as revenue from oil and gas privatization decreases over time, elite competition over private and public funding sources is likely to intensify. This will manifest itself, for example, in increasing battles between central and regional authorities over the supervision of foreign companies operating in the oil and gas regions of western Kazakhstan and over shares in the proceeds from foreign investment. Second, corruption is likely to continue and intensify for these reasons. Over the long-run, this will serve to impede democratization and hinder economic growth. The links between economic decline and corruption are well-documented and widely recognized. In the post-Soviet context, however, corruption—or, more precisely, so-called government drives against corruption—are likely to be used to justify the continued use of anti-democratic measures and the centralization of power.[96] Finally, the concentration of power in the central government will continue both to increase tensions between central and regional elites and to impede the development of more decentralized forms of political and economic decision making. Such centralization is also strongly associated with low levels of economic growth.[97]

COMBINED EFFECTS: INCREASING SOCIOECONOMIC PROBLEMS AND POLITICAL UNREST

Overly optimistic expectations of oil and gas wealth, combined with an increasing reliance on foreign investment, first and foremost, have greatly diminished the government's commitment to direct social spending. Spending in relation to GDP fell from 11.2 percent in 1992 to 6.6 percent four years later, benefit levels have declined substantially in real terms, and pensions are less than 50 percent of 1993 levels.[98] The results include increased rates of poverty, health problems, and epidemics. From 1996 to 1998, the gap between Kazakhstan's wealthiest and poorest citizens "widened dramatically," and today, according to the Red Cross, "three-quarters of Kazakhstan's 15.7 million population lives below the poverty line."[99] The number of people infected with, and deaths from, tuberculosis have also risen rapidly since 1996. In 1998, "the incidence of tuberculosis in Kazakhstan [was] about seven times that in the West, with at least 50,000 people, or 67 out of every 100,000, suffering from the disease."[100] Western aid organizations and local medical professionals alike cite the lack of resources for treating patients, as well as poor nutrition and bad hygienic conditions stemming from the sharp decline in living standards since independence as the primary causes.[101] It is not difficult to imagine the long-term consequences of these trends: the spread of disease throughout Kazakhstan and into neighboring states, the debilitation of an entire population due to poor health, and spiraling health care costs in a country with extremely limited fiscal resources.

A second related effect is the emergence of cycles of protest, crackdown, and co-optation in Kazakhstan over the past few years. The government's initial strategy of utilizing prospects of energy wealth to diffuse popular discontent has instead led to increasing social unrest. The official response, however, has been to threaten, punish, or attempt to co-opt opposition leaders rather than to address Kazakhstan's mounting social and economic problems. Since 1996, the emerging pattern is for the government to ignore protestors' demands as long as possible, to make modest concessions (if any at all), and to punish their leaders. Opposition leaders are often induced to accept positions within the government in lieu of continued police harassment and in exchange for their silence.[102] The series of strikes and standoffs that took place in October and November 1997 is a case in point. Demonstrators were initially warned, then ignored, and then physically prevented from carrying out their planned march. Those involved in leading the strikes, particular union leaders, were either punished with jail sentences and fines or "offered" an official governmental post.

Over the long-term, the continuation of these cycles may lead to the radicalization and regionalization of the opposition. As the government bureaucracy continues to shrink under Nazarbaev's streamlining program, for example, he will have fewer positions with which to buy off the opposition's leaders. Similarly, as regional socioeconomic differentiation increases and competition between central and regional elites over spheres of influence intensifies, the opposition might find it in its best interest to direct its grievances toward regional leaders.

CONCLUSION

The harsh scenario depicted here is not inevitable. There are several significant steps that the Kazakhstani leadership and/or the United States government can take in order to avert what appears to be an impending crisis. First, the Kazakhstani government should prepare itself for a moderate and protracted flow of oil revenues rather than a gusher. It must eliminate the widespread perception that energy wealth and foreign investment can act as a panacea for the country's social and economic ills. Government officials should focus their efforts instead on securing domestic sources of present and future revenue streams and securing investment to develop them. These efforts should also include building state institutions that have the capacity to prescribe and collect taxes.

Second, the United States should emphasize building an internal pipeline system in Kazakhstan for bringing oil and gas to the domestic market. Washington's focus to date has been on securing regional support and foreign investment for the construction of external pipelines to bring Kazakhstan's oil and gas to foreign markets. Yet an internal system is also key to Kazakhstan's future development. This would not only decrease Kazakhstan's reliance on Russia and Uzbekistan for energy supplies and its need to barter commodities for energy, but also would lay the foundation for broadening domestic industrial development.

Third, both the Kazakhstani and U.S. governments should encourage greater development in the manufacturing and agricultural sectors. This is key to achieving a healthy economy (and avoiding Dutch Disease), as well as to reducing poverty levels, both of which require broad-based growth.[103] In contrast to Azerbaijan, to which the Soviet legacy bequeathed little economic potential aside from its oil reserves, Kazakhstan inherited an industrial base in its north and east and produced a substantial portion of Soviet grain.[104] This provides grounds for some optimism. Yet both its manufacturing and agricultural sectors are severely constrained by huge debts, dilapidated equipment, formidable export obstacles, and poor environmental conditions. In the short term, the first two of these constraints might be mitigated through privatization, which Kazakhstan has already pursued in its manufacturing sector but not in agriculture.[105] Kazakhstan's strategy also fails to address several long-term developmental issues, including the country's collapsed educational and health care system and the state's weakening institutional capacity to provide such services. These are the key areas toward which the Kazakhstani leadership should turn its present attention and the U.S. government should direct any future aid. One of Kazakhstan's first steps should be to alter its privatization strategy to stress direct taxation of foreign companies so that the central and regional governments can develop the institutional capacity to build (or rebuild) local infrastructure on their own. The encouragement of open competition among domestic entrepreneurs for the opportunity to perform these functions themselves should follow once transparency in the tender process can be guaranteed.

Fourth, and closely related, Kazakhstan should resist the temptation to close its borders to free trade. Recent research demonstrates that those countries rich in natural resources that have maintained a commitment to free trade have also been the most successful economically—because open borders encouraged them to develop competitive manufacturing sectors.[106]

Finally, and perhaps most importantly, if Kazakhstan is to address all of its underlying social and economic problems, it must reopen its political space by allowing public dissent, freedom of the press, and truly competitive elections. Democratization is the key to ensuring that Kazakhstan's citizens can hold their leaders accountable for the way in which oil and gas revenues are used. While this is no guarantee that the use of oil and gas revenues in Kazakhstan will be economically efficient, it will decrease corruption by making financial dealings in the energy sector more transparent. As the findings in Terry Lynn Karl's chapter suggest, the need for transparency can provide a sufficient incentive for government officials to separate their private interests from their public office and to strengthen central institutions responsible for collecting and redistributing state revenue. Norway, the one petroleum-rich country that was able to avoid the common pitfalls of state-led development, was already a full-fledged democracy when it started to accrue oil rents.[107] The U.S. government's role is particularly crucial to democratic development in Kazakhstan. While it has certainly paid lip service to democracy since Kazakhstan became independent, the United States has not effectively discouraged this country's descent into authoritarianism since 1995. In fact, its drive to create a secure environment for foreign investment in Kazakhstan's oil and gas sector, combined with its visible support of other authoritarian leaders in neighboring resource-rich states, may have actually encouraged this trend. Thus, the United States could also help to steer Kazakhstan toward a more democratic form of government by structuring economic and political incentives accordingly. Such incentives include restricting future aid as well as future visits to the White House unless opposition parties and politicians are allowed to develop and participate in the political process unhindered, media censorship and the intimidation of journalists are eliminated, and international human rights norms are consistently observed.

NOTES

I am grateful to Mark Brzezinski, Minh Luong, Terry Lynn Karl, Rajan Menon, and Erika Weinthal for their comments on an earlier draft, and to Harold Collet for his research assistance. All interviews cited in this chapter were conducted with Erika Weinthal and made possible due to the generous support of two Collaborative Grants in Sectoral Policy provided by the National Academy of Sciences, National Research Council, Office for Central Europe and Eurasia, Washington, D.C. Responsibility for the contents and analysis, however, is mine alone.

1. N. A. Nazarbaev, "Kazakhstan—2030. Protsvetanie, bezopasnost' i uluchshenie blagosostoianiia vsekh kazakhstantsev," *Kazakhstanskaia pravda*, October 11, 1997, p. 15.

2. See, for example Richard Auty, "Does Kazakhstan Oil Wealth Help or Hinder the Transition?" *Development Discussion Paper*, no. 615 (Harvard Institute for International Development, December 1997), 3.

3. Author's interviews with USAID officials and consultants in Almaty, Kazakhstan, March and December 1997. Some oil industry experts have claimed that the U.S. government has also deliberately inflated the Caspian's potential reserves to attract more foreign investment there. See Patrick Crow, "Watching Government: Caspian Dreams," *Oil and Gas Journal Online* 96, no. 43 (October 26, 1998), <http://www.ogjonline.com/jet/> (November 7, 1998).

4. Auty, "Does Kazakhstan Oil Wealth Help or Hinder the Transition?" 14.

5. Pauline Jones Luong and Erika Weinthal, "Domestic Determinants of the International Role in Oil and Gas Development: The Case of Central Asia," unpublished paper, January 1999.

6. This is not to say that privatization itself is not a viable economic strategy, but that the form and manner in which it has been pursued in Kazakhstan produced unanticipated negative consequences. In fact, it may be the case that private ownership is the key to energy-rich states avoiding the "resource curse." See Michael Ross, "The Political Economy of the Resource Curse," *World Politics* 51 (January 1999): 296–322.

7. While members of the international community widely supported privatization and encouraged the Kazakhstani government to follow this economic strategy, they nonetheless expressed surprise and concern at the rate and degree to which Kazakhstan was privatizing its energy sector. Author's interview with Ruslan Mamishev, Operation Officer for Infrastructure, World Bank, Almaty, Kazakhstan, March 1997; and with Ken McNamara, USAID officer, Almaty, March 1997.

8. For details, see "Republic of Kazakhstan: Recent Economic Developments," *IMF Staff Country Report,* no. 98/84, August 1998.

9. For a fuller description of quasi- (or juridical) states, see Robert Jackson and Carl Rosberg, "Why Africa's Weak States Persist," *World Politics* (October 1992): 1–24.

10. Author's interview with a former Tengizchevroil executive in Almaty, Kazakhstan, December 1997. Chevron showed interest in Tengiz as early as 1988 and began negotiations with the Soviet government to develop these oil and gas fields shortly thereafter. Immediately after independence, Kazakhstan replaced Moscow as Chevron's counterpart in the negotiations.

11. Tengiz output rose from 20,000 b/d in 1993 to almost 200,000 b/d in 1998, with a projected target of 240,000 b/d in 2000. Additional production, however, is contingent upon the completion of the Caspian Pipeline Consortium's (CPC) planned pipeline from Tengiz to Russia's main Black Sea port at Novorossiisk. See Elshan Alekberov, "The Caspian and Central Asia: Despite Political Obstacles, Energy Work Progresses Around Caspian Sea," *Oil and Gas Journal Online* 96, no. 24 (June 15, 1998), <http://www.ogjonline.com/jet/> (November 7, 1998).

12. Author's interview with a former Tengizchevroil executive.

13. The Caspian Pipeline Consortium was established in 1993 to undertake major exploration projects in the north Caspian Sea. Drilling began in August 1999.

14. *OMRI Daily Digest,* no. 125, part 1 (June 27, 1996). In the past few years, however, several wells drilled in the south Caspian Sea have hit natural gas or come up dry, and oil companies have begun to question the basin's potential. See, for example, Steve Liesman, "Kazakhstan makes plans with China to Bolster its Oil Exports by 20%," *Wall Street Journal,* December 20, 1999.

15. *Pipeline News,* no. 41, part 2 (December 1996): 7–13, <http://www.intr.net/cpss/oil/journal.html> (December 20, 1996).

16. "A Central Asian Oil Dynamo Takes Shape," *Business Week,* October 27, 1997.

17. Bhamy V. Shenoy, S. Gurcan Gulen, and Michelle Michot Foss, "Caspian Oil Export Choices Clouded by Geopolitics, Project Economics," *Oil and Gas Journal Online* 97, no. 16 (April 1999): 3, <http://www.ogjonline.com.jet> (May 14, 1999).

18. "Kazakhstan Worried over Global Oil Price Falls," *Caspian Business Report* 2, no. 6 (March 30, 1998).

19. Gulbanu Abenova, "New Marketing Strategy for SHNOS," *The Globe Kazakhstan* 51, no. 271 (July 15, 1998); and *Dow Jones Newswires,* July 29, 1998. NB: Chevron, a western company, had a significant say in the decision to boost produc-

tion. The Kazakhstani government's inability to prevent this decision, if it so desired, is yet another consequence of its rapid privatization strategy.

20. *Energy & Politics* (formerly *Pipeline News*) 31, part 2 (September 17, 1998), <http://weber.u.washington.edu/~atyrau/EP/index.html> (September 17, 1998).

21. "Kazakhstan Produced 26 Million Tonnes of Oil in 1998," *Oil and Gas Journal Online*, January 19, 1999, <http://www.ogjonline.com/jet>.

22. For example, the International Energy Agency's most optimistic estimate is 1.54–2.34 million b/d by 2015, whereas Kazakhstani officials claimed they would export 2.6 million b/d by 2015. See: Shenoy, et al., "Caspian Oil Export Choices," 4.

23. *Pipeline News,* no. 47, February 8–14, 1997, <http://www.intr.net/cpss/oil/journal.html> (February 18, 1997).

24. *Kazakhstanskaia pravda,* November 28, 1997, p. 1.

25. "Four Candidates to Contest Kazakh Elections?" *RFE/RL Newsline* (November 25, 1998), <http://www.rferl.org/newsline/search/calendar1998.html>.

26. "A Central Asian Corporate Culture," *Business Week,* October 27, 1997.

27. "A New Capital is Being Built with Money from Sold Oil," *Focus Central Asia,* no. 3 (February 15, 1998): 8.

28. *Doroga k dorogoi Astane* (interview with Baltabek Kuandykov, president, National Oil and Gas Company, Kazakhoil) *Kazakhstanskaia pravda,* December 3, 1997, p. 1.

29. "Kazakhstan: Cold Comfort," *The Independent* (London), October 24, 1998, <http://www.eurasia.org.ru/english/november/Tnchk-an.html> (February 10, 1999).

30. Mark De Broeck and Kristina Kostial, "Output Decline in Transition: The Case of Kazakhstan," *IMF Working Paper* 69 (April 1998). See also *Delovaia nedelia,* June 14, 1996, p. 1.

31. "Republic of Kazakhstan: Recent Economic Developments," 44. NB: Thus far, Nazarbaev has only paid lip service to the idea of shifting focus to the manufacturing sector. See, for example, "Kazakhstan izmeniaet prioriteti" (Kazakhstan changes priorities), *Novoe pokolenie,* no. 14 (August 28, 1998).

32. Murat Asipov, "Kazakhstan: Living Standards During the Transition," *World Bank Report,* no. 17520-KZ (March 23, 1998): 7.

33. "Wage Debts in the Country Reached an Amount of 52 Billion Tenge: According to Trade Union Estimates, Social Tension is Growing in Kazakhstan," *Novoe pokolenie,* no. 18 (September 25, 1998). See also Asipov, "Kazakhstan: Living Standards During the Transition," 10.

34. Murat Asipov, "The Region Is Gone. During the Year, 300,000 People Emigrated from the Country," *Novoe pokolenie,* no. 17 (September 18, 1998).

35. It should be noted that government officials claimed in 1996 that out-migration rates were actually declining due to the country's "'increasing political and economic stability.'" See *OMRI Daily Digest,* no. 121, part 1 (June 21, 1996).

36. *Novaia gazeta,* March 17, 1998, 1.

37. *Novaia gazeta,* March 17, 1998, 1.

38. Auty, "Does Kazakhstan Oil Wealth Help or Hinder the Transition?" 11. See also Asipov, "Kazakhstan: Living Standards During the Transition," i–ii.

39. *RFE/RL Newsline,* July 23, 1997, and October 13, 1997, both at <http://www.rferl.org/newsline/search/calendar1997.html>.

40. "Republic of Kazakhstan: Recent Economic Developments," 20.

41. Auty, "Does Kazakhstan Oil Wealth Help or Hinder the Transition?" 5; and "The Next Shock?" *The Economist* (March 6, 1999): 23.

42. Auty, "Does Kazakhstan Oil Wealth Help or Hinder the Transition?" 5.

43. Auty, "Does Kazakhstan Oil Wealth Help or Hinder the Transition?" 6.

44. See *Non-OPEC Fact Sheet*, United States Energy Information Administration, March 1998, <http://www.eia.doe.gov/emeu/cabs/nonopec.html> (July 1, 1998).

45. Craig Charney, *Public Opinion in Kazakhstan 1996, Voices of the Electorate Series* (International Foundation for Electoral Systems: Washington, D.C.: April 1997), 17.

46. *Kazakhstan—The Public Speaks. An Analysis of National Public Opinion* (International Foundation for Electoral Systems: Washington, D.C.: September 1995), 21–22; and *RFE/RL Newsline*, July 28, 1997, <http://www.rferl.org/newsline/search/calendar1997.html>.

47. Pauline Jones Luong and Erika Weinthal, "The NGO Paradox: Democratic Goals and Non-Democratic Outcomes in Kazakhstan," *Europe-Asia Studies* (November 1999).

48. Auty, "Does Kazakhstan Oil Wealth Help or Hinder the Transition?" 14.

49. *Kazakhstanskaia pravda*, December 26, 1996, 1.

50. Asipov, "Kazakhstan: Living Standards During the Transition," 10; "Labour Protective Unions to Fight," *Focus Central Asia*, no. 3 (February 15, 1998): 34.

51. Kulyash Turgaziyeva, "Unemployed Population Has No Access to Medical Aid," *Delovaia nedelia*, no. 28 (July 24, 1998). See also: Asipov, "Kazakhstan: Living Standards During the Transition."

52. Asipov, "Kazakhstan: Living Standards During the Transition," 10; and Asipov, "The Region is Gone."

53. Asipov "Kazakhstan: Living Standards During the Transition," i–ii.

54. Asipov, "Kazakhstan: Living Standards During the Transition," 5.

55. Auty, "Does Kazakhstan Oil Wealth Help or Hinder the Transition?" 6.

56. Pauline Jones Luong and Erika Weinthal, "Energizing the Post-Soviet State: Energy Development Strategies in Post-Soviet Central Asia and Prospects for Democracy," unpublished paper, September 1998.

57. See, for example, Robert Dahl, *Polyarchy: Participation and Opposition* (New Haven, Conn.: Yale University Press, 1971); and David Held, *Models of Democracy* (Stanford, Calif.: Stanford University Press, 1987).

58. For a more sophisticated theoretical treatment of this link, see Douglas North and Barry Weingast, "Constitutions and Commitment: The Evolution of Institutions Governing Public Choice in Seventeenth-Century England," *The Journal of Economic History* 49 (1979): 803–33; and Robert Bates and Da-Hsing Donald Lien, "A Note on Taxation, Development, and Representative Government," *Politics and Society*, 14 (1985): 53–70.

59. See, for example, Terry Lynn Karl, *The Paradox of Plenty: Oil Booms and Petro-States* (Berkeley: University of California Press, 1997), 61–62.

60. *RFE/RL Newsline*, May 29, 1997, <http://www.rferl.org/newsline/search/calendar1997.html>.

61. Asipov, "Kazakhstan: Living Standards During the Transition," 4.

62. *Pipeline News*, no. 46 (February 1–6, 1997), <http://www.intr.net/cpss/oil/journal.html> (February 10, 1997); Author's interview with Ruslan Mamishev, Operation

Officer for Infrastructure, World Bank, Almaty, Kazakhstan, March 1997; and Chris Bird, "Investors eye Kazakh Privatisation with Caution," *Reuters,* December 17, 1996.

63. Author's interviews with oil and gas executives working in Kazakhstan; Almaty, December 1997.

64. *Energy & Politics,* no. 32, part 2 (September 28, 1998), <http://weber.u.washington.edu/~atyrau/EP/index.html> (September 28, 1998). NB: OKIOC was created to develop Kazakhstan's projected reserves in the Caspian Sea.

65. Author's interviews with USAID consultants and oil and gas executives working in Kazakhstan; Almaty, March and December 1997. As David Hoffman reports, even the Azerbaijani government will realize only about 80 percent of total revenues from what is considered its most lucrative production-sharing agreement, the PSA with the Azerbaijan International Operating Company. See chapter 4, "Azerbaijan: The Politicization of Oil," by David Hoffman, in this volume.

66. *Pipeline News,* no. 46 (February 1–6, 1997).

67. Author's interviews with oil and gas executives working in Kazakhstan; Almaty, March and December 1997. See also *Pipeline News,* no. 9 (April 13–19, 1996), <http://www.intr.net/cpss/oil/journal.html> (April 24, 1996); and *RFE/RL Newsline,* September 12, 1997, <http://www.rferl.org/newsline/search/calendar1997.html>

68. Author's interviews with oil and gas executives working in Kazakhstan; Almaty, March and December 1997. See also Hugh Pope, "To Get Kazakhstani Oil, Small Firm Takes on Soccer Team, Camels: At Hurricane Hydrocarbons, Officials Find They Fill a Big Post-Soviet Role," *Wall Street Journal,* November 18, 1997, 2.

69. Merhat Sharipzhan, "Kazakhstan: Workers' Protest Reveals Flaws in Privatization," *RFE/RL Newsline,* November 14, 1997, <http://wwwrferl.org./nca/features/1997/11/F.RU.97.111413328.html> (December 18, 1998). NB: Privatization was in fact a common government strategy across sectors for "solving" the problem of wage arrears. See, for example, Alastair MacDonald, "Kazakh Steel Town Struggles Back to Life," *Reuters,* March 23, 1998; and Merhat Sharipzhan, "Kazakhstan: A Case Study Of Privatization Problems," *RFE/RL Newsline,* January 7, 1997, <http://www.rferl.org/nca/features/1997/01/F.RU.970107171240.html> (December 18, 1998).

70. Brigit Brauer, "Chevron's Kazak Oilfield Wealthy," AP Newswire, July 17, 1998.

71. "Labour Protective Unions to Fight Against Reduction," *Focus Central Asia,* no. 3 (February 15, 1998): 34.

72. See, for example, *Energy & Politics* 27, part 2 (August 11, 1998), <http://weber.u.washington.edu/~atyrau/EP/index.html> (August 11, 1998).

73. "Tractebel to Promote a New Prime Minister?" *Focus Central Asia,* no. 4 (February 28, 1998): 7–10. NB: Almatyenergo is the main provider of both electricity and heat to Almaty.

74. Merhat Sharipzhan, "Kazakhstan: Workers' Protest Reveals Flaws in Privatization."

75. "Kazakhstan Aspires to the Sea," *Focus Central Asia,* no. 3 (February 15, 1998): 27.

76. As quoted in *Reuters Newswire,* April 21, 1997.

77. "Tractebel to Promote a New Prime Minister?" *Focus Central Asia,* no. 4: pp. 7–10.

78. *Pipeline News,* no. 46 (February 1–6, 1997). See also "Kazakh Oil Barons, Government, Clash on Energy Sell-off," *Reuters Newswire,* February 3, 1997; and *Pipeline News,* no. 47 (February 8–14, 1997).

79. *Pipeline News*, no. 47 (February 8–14, 1997).

80. Author's interviews with foreign oil and gas executives working in Kazakhstan, March and December 1997.

81. See Jones Luong and Weinthal, "Energizing the Post-Soviet State."

82. *Karavan*, April 10, 1998, p. 1.

83. "The Struggle with Corruption: Can a Snake Bite its Own Tail?" *Focus Central Asia*, no. 3 (February 15, 1998); and *RFE/RL Newsline*, July 28, 1997.

84. See Jones Luong and Weinthal, "Energizing the Post-Soviet State."

85. As quoted in "Kazakhstan Aspires to the Sea," 26.

86. For example, most foreign observers (including OSCE representatives) declared the January 1999 presidential elections "neither free nor fair," and political party identification remains extremely low. See "End Note," *RFE/RL Newsline*, May 7, 1999; and Charney, "Public Opinion in Kazakhstan."

87. Asipov, "Kazakhstan: Living Standards During the Transition," 15.

88. Asipov, "Kazakhstan: Living Standards During the Transition," 17.

89. See Ralph S. Clem, "Interethnic Relations at the Republican Level: The Example of Kazakhstan," *Post-Soviet Geography* 34, no. 4 (April 1993): 231. It should be noted here that there was a sharp decline in the Kazakh population in the 1930s due to the collectivization of agriculture under Stalin, which amounted to forced sedentarization and the extermination of the Kazakhs' traditional way of life.

90. Clem, "Interethnic Relations at the Republican Level," 231. These factors include the historical settlement of Russians in what is today northern and eastern Kazakhstan, and several key Soviet policies, such as the incorporation of lands comprised primarily of Russians into the northeastern part of the Kazakh Soviet Socialist Republic and the deliberate promotion of Russian emigration into northeastern Kazakhstan to modernize the economy. For details, see Geoffrey Wheeler, *The Modern History of Soviet Central Asia* (New York and Washington, D.C.: Frederik A. Praeger, 1964), 76–77; C. N. Pokrovskii, *Obrazovanie Kazakhskoi ASSR* (Alma-Ata, Kazakhstan: Academy of Sciences, 1951); and V. K. Savos'ko, *Preobrazovanie Kazakhskoi ASSR v Soiuznuiu Respubliku* (Alma-Ata, Kazakhstan: Academy of Sciences, 1951).

91. For details, see Pauline Jones Luong, *Designing Democracy: Institutional Continuity and Change in Transitional States*, unpublished manuscript, 1998, chapter 4.

92. See Sharipzhan, "Kazakhstan: A Case Study Of Privatization Problems."

93. Sharipzhan, "Kazakhstan: Workers' Protest Reveals Flaws in Privatization."

94. Merhat Sharipzhan, "Kazakhstan: Labor Unrest Has Roots In Failed Privatization," *RFE/RL Newsline*, March 12, 1998, <http://search.rferl.org/nca/features/1998/07/F.RU.980713125744.html>. "Union Urges Government to Help Workers in Mangistau," *Energy & Politics* 27, part 2 (August 11, 1998), <http://weber.u.washington.edu/~atyrau/EP/index.html>.

95. Jones Luong and Weinthal, "Energizing the Post-Soviet State."

96. Paul Goble, "Kazakhstan: Analysis From Washington—The Corruption Of Power," July 13, 1998, <http://search.rferl.org/nca/features/1998/07/F.RU. 980713-125744.html>.

97. See Philip Lane and Aaron Tornell, "Power Concentration and Growth," *Discussion Paper*, no. 1720 (Harvard Institute of Economic Research, May 1995).

98. Asipov, "Kazakhstan: Living Standards During the Transition," iii.

99. "Kazakhstan: Cold Comfort," 1.

100. *OMRI Daily Digest,* no. 64, part 1 (March 29, 1996).

101. *ITAR-TASS,* Moscow, December 13, 1998; and *OMRI Daily Digest,* no. 64, part 1 (March 29, 1996).

102. For a recent example, see "Opponent No. 1 Has Become An Official Employee," *Focus Central Asia,* no. 4 (February 28, 1998): 2–3.

103. For more on Dutch Disease and related problems, see chapter 3, "Crude Calculations: OPEC Lessons for the Caspian Region," by Terry Lynn Karl.

104. For details on Azerbaijan, see Hoffman, "Azerbaijan: The Politicization of Oil"

105. Export difficulties and poor environmental conditions are much more acute, particularly in the agricultural sector. For example, while Kazakhstan has found foreign markets for its steel in Southeast Asia and the Middle East, high transportation costs render the export of its grain unprofitable. This suggests that agriculture might best be developed for the domestic market, but this too is constrained by water shortages and polluted soil.

106. See, for example Jeffrey D. Sachs and Andrew M. Warner, "Natural Resource Abundance and Economic Growth," *Working Paper 5398,* National Bureau of Economic Research, December 1995.

107. See Karl, "Crude Calculations: OPEC Lessons for Caspian Leaders."

6

Turkmenistan's Energy: A Source of Wealth or Instability?

Nancy Lubin

The outwardly extravagant hotels that seem to emerge from nowhere in the desert outside of Turkmenistan's capital city, Ashgabat, remain a vivid reflection of the hopes and realities of Turkmenistan today. Built ostensibly to attract Western investment primarily in Turkmenistan's rich energy sector, today they are often devoid of energy executives. Instead, they are widely viewed as havens for money launderers, smugglers, and other criminal elements. Their fine wines and liqueurs seem to be shared among the hotel executives themselves, who often seem uninterested whether or not there are guests at their hotels to consume these luxury items.

In some ways, this image reflects Turkmenistan's hopes for a better future through the enormous wealth the country anticipates from exports of oil and natural gas. But it also reflects the more sinister, current realities of waste, crime, hardship, and potential instability that may engulf Turkmenistan well before energy revenues begin to flow.

What kind of wealth is Turkmenistan likely to see from its vast energy resources, and to what effect? What are the key sources of potential instability today—and to what extent is energy wealth likely to ameliorate or exacerbate them? What is the likelihood that potential energy revenues will benefit a broad spectrum of the population and encourage stable development for the country as a whole? And what are the key challenges for the United States as it balances important, but often competing, commercial, political, and security goals in this part of the world? These questions comprise the focus of this chapter.

TURKMENISTAN'S ENERGY:
REAL OR ELUSIVE HOPE FOR THE FUTURE?

Turkmenistan emerged as among the poorest and most authoritarian of the former Soviet republics when it achieved independence on October 27, 1991. At roughly 188,000 square miles, Turkmenistan is the fourth largest former Soviet republic. Yet its resources have seemed more limited, and its challenges more daunting, than those of other Caspian countries. Eighty percent of Turkmenistan is desert, and as primarily a raw material supplier for the former Soviet Union, the country was the least developed of all of the newly independent states. With the highest infant mortality rates in the former USSR, its sparse population[1] was also among the most destitute.

Hopes for Turkmenistan's development have been founded primarily in the enormous untapped energy wealth that lies beneath this desert. Although Turkmenistan's oil deposits, at about 1.2 billion barrels, are small relative to other Caspian countries, Turkmenistan is home to roughly 102 trillion cubic feet in gas reserves—ranking it third in the world after only Russia and Iran.

But the years since independence have been full of ups and downs in getting these resources to market. Before the breakup of the USSR, Turkmenistan's natural gas was exported through a single export pipeline that crossed Russia. Thus, following independence, gas exports remained under the control of Moscow. Since the mid-1990s, however, Russia has refused to export Turkmen gas to markets outside of the Commonwealth of Independent States (CIS); has kept prices low; and as countries within the CIS, particularly Ukraine, fell increasingly into debt, Turkmenistan frequently did not receive payment for deliveries. As a result, in March 1997 Turkmenistan effectively halted its gas exports to Russia. These factors have severely impacted Turkmenistan's economy.

Thus, while energy remains the mainstay of the Turkmen economy, its role has been rocky and uncertain. In December 1997, a small pipeline was opened to export a minor amount of gas from Turkmenistan to Iran (the only export route, however small, from the former Soviet republics that is independent of Russia). At the end of December 1998, Russia's Gazprom agreed to a deal allowing Turkmenistan to resume shipments of gas to Ukraine. But the amount of gas exported to Iran is marginal, and the resumption of gas shipments to Ukraine is expected to be less profitable than anticipated. The agreement with Ukraine, for example, calls for Turkmenistan to provide Ukraine with 20 billion cubic meters of natural gas in 1999, 40 percent of which will be paid for in hard currency and 60 percent in goods and services. Many observers believe, however, that the latter percentage will turn out to be far higher, and that the agreement will be almost entirely a barter deal.

During the past few years, Turkmenistan has explored a range of large scale pipeline options—through Afghanistan to Pakistan; through Uzbekistan to

China; and through Iran to Turkey—but none of these currently looks promising in the short term. Because of the ongoing civil war in Afghanistan, in August 1998 Unocal, the lead western company involved in the pipeline project through Afghanistan to Pakistan, suspended its participation, and in December of that year announced that it was pulling out altogether. Turkmenistan's hiring of Royal Dutch/Shell to explore pipeline options through Iran was nipped in the bud by U.S. pressure and support of an alternative route under the Caspian Sea and through the Caucasus to Turkey.

This latter route, the trans-Caspian pipeline, currently remains the most promising export route for Turkmenistan's gas. An agreement recently signed in Ashgabat provides for the construction of a roughly 1,200-kilometer pipeline under the Caspian Sea through Azerbaijan and Georgia and then to Erzurum, Turkey (bypassing both Iran and Russia). The construction cost is estimated at $2 billion, and the pipeline would have a capacity of upwards of 16 billion cubic meters of gas per year. In December 1998, Turkmenistan's president Saparmurat Niyazov reportedly said that he had decided to postpone indefinitely plans for construction of a gas export pipeline from Turkmenistan via Iran to Turkey out of preference for the alternative trans-Caspian route.[2]

But this pipeline, too, is no sure thing. While the United States, Azerbaijan, and Georgia back the project, Russia and Iran, among others, are strongly opposed to it. The Russian Foreign Ministry stated that the pipeline plans are "premature," and urged that the project be delayed until environmental impacts are carefully studied and the legal status of the Caspian Sea has been decided—i.e., perhaps indefinitely. Iran hopes to return to the Turkmenistan-Iran-Turkey option, with Turkmenistan's gas transported through its own territory. And both Iran and Russia object to a plan that is so explicitly supported and justified by the United States on political grounds—i.e., to avoid domination of Central Asia by either of those two regional powers.

This opposition, as well as a host of other factors, could well put the project on hold. According to some accounts, the project could be killed by current plans on the part of Gazprom and the Italian company ENI to pursue the so-called Blue Stream pipeline project to construct a 400-km gas pipeline, with an annual capacity of 16 billion cubic meters, from Russia under the Black Sea to the Turkish port of Samsun, and then on to Ankara.[3] And budgetary limitations will further complicate construction of the trans-Caspian pipeline. In particular, although the U.S. government is strongly promoting the pipeline, it is limited in the financial assistance it can provide—and perhaps further limited by the Greek-American and Armenian-American lobbies who oppose the whole initiative as one that would only strengthen their long-time adversaries, Turkey and Azerbaijan.

To tap export potential, Turkmenistan has also been focusing on domestic development of the energy industry—modernizing its refineries and other facilities, encouraging foreign investment to develop production and pipelines throughout the country; and working to resolve disputes with other countries, particularly Azerbaijan, over ownership of offshore resources in the Caspian Sea, environmental standards, etc. But these efforts, too, present a mixed picture, and the future is unclear. While western energy companies continue to explore new investments in Turkmenistan, for example, three major western oil companies—Mobil (U.S.), Monument Oil (Britain), and Dragon Oil (Ireland)—are reportedly cutting the volume of work by half at deposits in Turkmenistan.[4] Smaller companies recently have done likewise.

As the energy picture unfolds, the development of Turkmenistan's economy has been on hold, with its future unclear. Production in other sectors has not filled the gap. The 1998 cotton harvest—Turkmenistan's other major export—was disappointing, reportedly totaling less than half of planned targets. In an address before Turkmenistan's State Council in December 1998, President Niyazov expressed great disappointment and lowered plan targets for 1999.[5] Production of grain was likewise disappointing: although the Turkmen government claimed that the 1998 grain harvest exceeded targets, flour is increasingly scarce; and the price of government-subsidized flour—for the poor and families with many children—has reportedly quadrupled since November 1, 1998, from 25 to 100 manat per kilogram.[6] In short, with so much of Turkmenistan's development reliant on energy, the country has declined into economic hardship, exacerbating already enormous and potentially destabilizing pressures that have only grown in its wake.

POTENTIAL SOURCES OF INSTABILITY

These economic pressures and the general drop in living standards currently comprise perhaps the most important source of tension and potential instability in Turkmenistan—both because of the growing poverty and also because of the enormous and highly visible gap between rich and poor that is only widening. Indeed, Turkmenistan's general economic situation has deteriorated dramatically since the early 1990s when hard currency revenues for the country's natural gas exports began to dry up. Real gross domestic product reportedly declined more than 65 percent between 1993 and 1997,[7] falling by over 27 percent in 1997 alone. And 1998 saw little improvement. Most of Turkmenistan's basic infrastructure—including its roads and other transportation systems; its water, gas, and electric power distribution systems; telecommunications; and other facilities—dates from the Soviet period and is crumbling. Although water, gas, and electricity have been provided to the Turkmen population free of charge since January 1993, supplies are limited, if available at all. Water is available for apartments even in the capital city only a few hours per day, and all of Turkmenistan experiences frequent power outages.

Wages in Turkmenistan likewise remain low. The average monthly wage in the first two months of 1998 was 195,000 manat[8] (at the time, $46 at the official exchange rate, and about one-third that at the market rate). Indeed, an extraordinarily high proportion of income is spent on food—and that is when one is paid. Turkmenistan's workers are often not paid their wages, or are paid so late that inflation erodes the value of their incomes before they even receive them. Although officially there is no unemployment, in fact unemployment and under-employment are high.

The impact of the deteriorating economy on the health and well-being of the population has been enormous. While official government statistics suggest that infant mortality (often regarded as the main indicator of the health of a population), has declined since independence to about 73 deaths per 1,000 live births as of 1997, local specialists believe it remains extremely high, and at least twice officially reported levels.[9] Doctors have told me that if they were to report actual numbers of infant deaths, they would lose their jobs.[10] And as the population's incomes and living standards sharply decline, medical services and medicines have become only more expensive, viewed as luxuries available only for the privileged few. Turkmenistan reportedly has the lowest life expectancy of any of the former Soviet republics.

These strains are only exacerbated by the serious decline in social services across the board. Reportedly, pensions are frequently unpaid; health care has deteriorated just as it has become unaffordable; and public education has likewise declined seriously as school facilities deteriorate, as families keep their children home from school to help make ends meet, and as teachers seek other sources of income outside of the school (working on their private plots or seeking other additional work) to supplement the meager roughly $20 a month they receive in wages (when they are paid). Schools have closed as the number of school-aged children continues to grow.

Declines in agricultural production; nonpayment of wages, particularly in agriculture in which an estimated 50–70 percent of the population works; the fall in the value of the currency against the dollar—all exacerbate a desperate situation for many Turkmen citizens. While the resplendent presidential palace, new stadiums, and massive monuments estimated to cost hundreds of millions of dollars—including the roughly 300-foot-high Monument of Independence—are intended to inspire the population with visions of the glorious future that awaits Turkmenistan, they have also increasingly become symbols of the dramatic gap between the wealth held by some and the population's growing destitution.

These socioeconomic strains are only exacerbated by competition over scarce resources and growing environmental problems; by growing crime and organized crime; and by Turkmenistan's authoritarian political environment. Severely limited water resources, for example—and exceedingly poor water management—have led to strains not only within Turkmenistan, but between

Turkmen citizens and their Uzbek neighbors. The Amu Darya River, the main source of water that runs along the border of these two countries, has already been the scene of conflict between Uzbek and Turkmen farmers who believe their very livelihood, if not the very the survival of their families, may be at stake. The Aral Sea has likewise become a deepening source of tension. Once the world's fourth largest inland body of water, the sea has shrunk to half its 1960 area, and one-third of its 1960 volume, leaving devastating environmental impacts in its wake. And these issues are but the tip of the iceberg in Turkmenistan, where a whole range of environmental problems[11] also seriously affects the population's health and living standards today and represents imposing obstacles to future economic growth.[12]

Sharing a 744 km border with Afghanistan, a 992-km border with Iran, a 1,621-km border with Uzbekistan, and a 379-km border with Kazakhstan, Turkmenistan is also a regional transit route for narcotics, arms, and reportedly weapons of mass destruction. Turkmenistan is a key staging point for narcotics traffickers to move opium, heroin, and other illicit drugs from the East to the West, and to smuggle precursor chemicals to Southwest Asia. Opium from Afghanistan, Uzbekistan, Tajikistan, and Iran transits Turkmenistan to markets in Turkey, Russia, and the West—a trade which reportedly has been exacerbated by expanding air links and the railway line between Iran and Turkmenistan that officially opened in 1996.[13] Cargo traffic on the railway and on trucks is reportedly quite heavy, carrying drugs and arms into neighboring countries. And money laundering is a natural by-product, growing particularly through the burgeoning number of luxury hotels and casinos.

Officials are also concerned that drug cultivation and use within Turkmenistan itself may be growing. Domestic cultivation, while illegal, occurs mostly along the Iranian border in the Ahal region (which includes Ashgabat) and in the eastern parts of the country (Lebap and Mary regions). Although health authorities indicate that domestic drug use currently is not a major problem, others believe the numbers may be growing and will continue to grow as smuggling expands, as local production grows, as living standards deteriorate, and perhaps as Turkmenistan's own refugee population grows as a result of instability and war in neighboring countries.

Finally, the ethnic, tribal, and political environments of Turkmenistan may be giving rise to tension and potential instability in their own right. With roughly a quarter of Turkmenistan's population comprised of non-Turkmen ethnic groups,[14] and with a mix of tribal ties that go back for centuries, Turkmen society is shaped by a delicate balancing act. It is difficult to assess the precise impact of these elements on Turkmen society: some argue that these divisions remain strong, divisive undercurrents throughout Turkmen society; others, that tribal consciousness has been greatly diluted over the years and poses little, if any, threat to stability at home; and still others believe that while significant, the potency of tribal ties is limited to certain segments of the population—i.e., that it is a factor of age, gender, or rural/urban distribution.

While the extent to which ethnic rivalries and tribal divisions may be sources of conflict in their own right is debatable, they do remain potent umbrellas under which other grievances could well be expressed. President Niyazov has referred to potential strife between clans and tribes in speeches throughout the 1990s, and has sought to defuse any potential strife through, for example, personnel policies and symbolic measures, such as the design of the Turkmen flag itself. Nevertheless, tribal and clan divisions could well become important fault lines should economic or other pressures among the population intensify.

Perhaps most threatening to stability is Turkmenistan's own human rights record. Turkmenistan's record on human rights abuses is considered among the worst of any former Soviet republic by Freedom House, Human Rights Watch, and the U.S. Department of State. Any public expression of grievances is quickly suppressed, and the number of human rights abuses remains high. Not only is this a serious source of dissatisfaction in its own right; but the mere fact that there is no outlet for any kind of discontent can politicize an otherwise quiescent population.

Although political opposition is severely muted in Turkmenistan, this confluence of tensions and strains has already triggered unrest, most notably in July 1995, when a large, unprecedented, and well-organized anti-government protest erupted in Ashgabat. An estimated one thousand citizens—although at the time reports on the street suggested several thousands—gathered on the city's main street complaining of hardship, criticizing the disconnect between the leader's palaces and their own destitution, and calling for the removal of the president. The demonstrators quickly dispersed when about 80 were taken into police custody. For most of us in Central Asia at the time, the reports of numbers, catalysts, and official reactions remained confused and contradictory; they most certainly seemed that way to Turkmenistan's citizens as well. Turkmen authorities alleged that the demonstrators had been provoked by people high on drugs and alcohol, and that political protest was still absent from Turkmenistan. But the demonstrations highlighted the enormous tensions beneath the surface of what had seemingly been the most quiescent of Central Asia's populations. As noted above, living standards have only continued to deteriorate since those demonstrations.

CURRENT POLICY IN TURKMENISTAN

Paramount within Turkmenistan's leadership seems to be the necessity to maintain control and stability at home.[15] Formal and informal statements by government leaders suggest that Turkmenistan's game plan thus far has been largely to buy time—i.e., to continue to provide subsidies for basic goods and services to the population that might help avoid the kinds of conflict that have emerged elsewhere in the former USSR; to initiate the trappings of reform, while maintaining strict control over all aspects of economic and political life; and to work to increase revenues, particularly from the export of gas on the world market, so as to make both of the above feasible.

This approach has led to a situation whereby whatever reforms the government has given with one hand, it has negated or countered with the other. For example, "privatization" has become quite a buzzword in Ashgabat; but while the government has initiated a number of privatization policies, it has not implemented the changes to make those policies viable. Instead, all decisions continue to be made at the highest levels of government; economic decisions continue to be influenced strongly by political considerations; the legal foundations of a market economy remain weak or absent; and privatized businesses continue to be impeded by arbitrary and changing regulations, tax laws, and the like.

In a still centrally-planned economy, the government for the most part continues to control all inputs, pricing, transportation, and access to international markets. And to ensure that social hardship is kept to a minimum, the government also has the right to control employment even in privatized enterprises.[16] Indeed, because of these and other obstacles and difficulties in finding accessible and viable financing, the privatization rate has reportedly slowed considerably. The government of Turkmenistan has also created free economic zones, but except for Ashgabat's airport zone, these, too, have been slow to become operational.

Turkmenistan has also put a high priority on attracting foreign investment, not only in energy but in all sectors of the economy. But this, too, has been greatly inhibited by similar obstacles.[17] For example, President Niyazov still makes all final decisions in contract awards, which continue to be granted for political reasons "without regard for feasibility, economic viability and/or the ability of the winning company to do the job." There is no effective legal system in place to enforce rules and regulations. And issues taken for granted in the West, such as the sanctity of contracts, are by no means assured by a government with quite a different track record.[18]

In general, the full range of economic reforms in Turkmenistan—through decrees, laws, administrative rules, and the like—is often internally contradictory; decrees and regulations are frequently revised, with changes applied retroactively; and laws are often implemented arbitrarily or not at all. While intended to encourage foreign investment and improve performance at home, the policies of the Turkmen government may be doing just the opposite—inhibiting investment and raising doubts as to whether energy revenues will be a blessing or a curse. All the while, economic hardship increases.

President Niyazov has stated that despite any economic setbacks, the 1999 budget of about 20.5 billion manat (about $4 billion in early 1999) is sufficient to maintain health care at current levels, retain subsidies on flour prices, and provide free gas, water, energy, and salt to the population as a whole.[19] He continues to assert that Turkmenistan is committed to "measured, but effective steps to create a market-oriented economy and improve the daily life of the Turkmen people," but only by avoiding conflict and strife. Yet western economists believe these priorities are severely misplaced, and that this policy is only causing more hardship, aggravating "an already serious economic crisis," and "causing the worst-off people to suffer the most."[20]

Likewise on the political side, since independence Turkmenistan has adopted many of the trappings of a democratic system. According to the constitution, adopted in May 1992, Turkmenistan is a democratic state—a "presidential republic" governed by the rule of law and with a series of checks and balances inherent in the institutions of the executive, legislative, and judicial branches. But the president of the republic continues to exert such complete control over all of these institutions—for example, through his power to appoint and dismiss all key actors throughout all three branches of government, to dissolve the Parliament and other institutions, and to issue decrees and orders that have the power of law—that all power effectively resides in the hands of President Niyazov alone. The president denies the existence of any opposition, and the need for additional parties outside of his own. This is only reinforced by the "cult of personality" that has emerged around the president— or "Turkmenbashi" (the head of the Turkmen people), as he prefers to be called—since independence. The disproportionate amounts of money spent on grandiose projects are all part of this cult of personality, glorifying, in the words of one local, a country and a leader that are one and the same.

One area where there has been little even in the way of rhetorical policy movement regards narcotics trafficking and other aspects of corruption and crime. While President Niyazov has conducted periodic house-cleanings and crackdowns on corruption among personnel at the middle and highest levels of government, there reportedly has been no movement on plans to formulate a national drug policy or to develop institutional structures to begin to address this problem. Niyazov has frequently accused law enforcement bodies and other high officials of collusion in drug smuggling—and emphasized that this problem extends to "practically every [law enforcement and military] organ"[21]—but little has been done. The government of Turkmenistan has likewise been reluctant to expand drug treatment programs, reportedly for fear that this will create greater interest in drug use.

Despite the intolerance for opposition, however, local nongovernmental organizations have emerged in Turkmenistan that provide some outlet for community concerns and may ultimately have far greater impact throughout Turkmenistani society. Groups such as the Ashgabat Environment Club Catena, for example, founded in 1994 as a community organization concerned with environmental problems, have developed into key providers of information—in this case linking scientists and others through e-mail, information data bases, clubs, and ties to international organizations. This particular club hopes to create independent sources of information for all groups of society; to help establish public environmental control; to promote new methods of environmental management; and even to encourage industrial development based on sound environmental assessments. While no organizations in Turkmenistan are independent of government control, efforts such as these may help create a base for greater public awareness and participation in the future.

In the meantime, a glossy brochure distributed on the occasion of President Niyazov's visit to the United States in April 1998 boasts, "Under the policies of President Saparmurat Niyazov, Turkmenistan is among the most stable of the former Soviet states." But this stability is based on fear and quiescence, and on an implicit understanding that the Turkmenistan government will continue to provide basic goods and services at nominal prices or free of charge—a strategy increasingly brought into question. This strategy could well backfire, bringing all of these tensions to the fore long before energy revenues begin to flow.

IMPACT OF POTENTIAL ENERGY WEALTH ON DEVELOPMENT AND STABILITY

Turkmenistan's track record, then, does not bode well should the country take in the expected billions of dollars each year from energy exports in the future. If current policies and priorities are any indication, there is little to suggest that needed steps would be taken to encourage sustainable economic development, benefit a broad spectrum of the population, and put Turkmenistan on a solid and stable economic and political footing in the early part of the next century.

That said, however, many questions remain. Given the complicated relations with Turkmenistan's neighbors, how great can Turkmenistan's potential "windfall" be, and over what time frame? Regardless of route, will Turkmenistan's future continue to be hostage to its neighbors—whether Russia, Iran, or Azerbaijan—for its revenues? If U.S. relations soften with Iran, as recent events suggest they could, will this significantly change potential routes and the picture for Turkmenistan's competitiveness?

How, in fact, will future gas revenues be used? How much can we extrapolate from today's policies, when they depend exclusively on the decisions of one man who himself is quite ill? President Niyazov, now 59 years old, was hospitalized in 1993 for lung surgery; in 1998, he underwent a quadruple bypass heart operation. With no obvious successor, will the passage of Niyazov—even if many years in the future—leave the door open to reformers to take the helm or lead to further authoritarian rule? And with no mechanisms for succession,[22] is the transition itself likely to be smooth, or a source of enormous upheaval and conflict?

Finally, as the chapters by Terry Lynn Karl, David Hoffman, and Pauline Jones Luong point out, sudden riches can be dangerous. High revenues from oil and gas have made some countries prosperous, but have helped destroy others. Turkmenistan's record to date, and the reluctance of Turkmenistan to institute needed fundamental reform, suggest the likelihood that Turkmenistan "will decline more into Nigeria-style chaos than rise to Norway-like stability and affluence"[23]—or perhaps at best that it will continue along the same lines, where rights continue to be denied, poverty abounds, but turmoil does not become strong enough to shake the foundations of the political order. But should more reform-oriented leaders take hold, could fundamental reform also trigger instability in Turkmenistan—as current leaders would argue—opening an impoverished country with no recent history of open debate to chaos and increased despair?

IMPLICATIONS FOR U.S. POLICY

These uncertainties create both opportunities and dangers for the United States. Turkmenistan's leaders state that the country's legitimacy on the international stage is advanced by: (1) their doctrine of "permanent neutrality" in foreign policy; (2) their efforts to shore up Turkmenistan's sovereignty and maintain independence from Russia, and from the CIS as a whole; (3) their contribution to regional stability, including their current role in hosting talks between the Taliban and Afghanistan's opposition; and (4) their country's vast commercial possibilities. These efforts, they claim, should be of particular interest to the United States. Turkmenistan has joined the ranks of other Caspian countries in paying U.S. lobbying firms—in their case, through the Israeli company Merhav—to keep tabs on U.S. Caspian policy and to "raise the visibility of Turkmenistan in the United States."

The United States, for its part, has stressed a number of competing interests, at least in the short term, that make the formulation of Turkmenistan policy quite difficult and raise serious questions for the future. A chief focus of U.S. policy has been tapping Turkmenistan's energy resources—as with the Caspian as a whole—both to diversify world energy supplies and to provide an opportunity for U.S. trade and investment. Another focus has been to encourage stability and independence in a country lying at the intersection of Russia, Iran, China, IndiaPakistan, and Afghanistan—all of which likewise raise serious and complex foreign policy questions for the United States.

Promotion of human rights, democratic and economic reform, and cooperation in combating smuggling and the proliferation of weapons of mass destruction in this part of the world, are key components of U.S. policy in Turkmenistan, as is encouraging the role of Turkmenistan in promoting prosperity and stability throughout Central Asia as a whole. But implementing these policies has been difficult in a country that does not share the same short-term goals and where human rights abuses remain high.

On his visit to the United States in April 1998, President Niyazov met with both President Clinton and Vice President Gore. Against the din of devastating media reports on Turkmenistan's poor human rights record, President Niyazov signed accords with U.S. firms on energy exploration and production. Despite the many obstacles, a number of U.S. companies view investment in Turkmenistan as promising.

But U.S. assistance policy remains complicated in Turkmenistan, raising serious challenges for the near future. To date, Turkmenistan has received among the smallest portions of U.S. assistance to the former Soviet Union, and in 1998 it received the least amount of U.S. assistance to any of the newly independent states.[24] This assistance consisted of an estimated $13.2 million, comprised of $5.29 million in Freedom Support Act funds, $4.95 million in Defense Department humanitarian commodities, and about $2.9 million in other U.S. government funds. Turkmenistan also benefited from regional initiatives in addressing oil and gas, water, and energy issues.

U.S. assistance in 1998 was spread thin among: (1) economic restructuring programs, including privatization support; (2) trade, investment, and business development programs; (3) democracy programs, including NGO development assistance; (4) energy and environmental programs; (5) medical and humanitarian programs; and (6) security programs, aimed both at increasing military ties and at providing counter-narcotics and other training to Turkmenistan's Border Guard Service and law enforcement community. There are also currently about sixty Peace Corps volunteers serving in Turkmenistan.

But reform programs are difficult to implement in a country where the leadership does not support those reforms, or, in the words of the U.S. State Department, in the face of a "failure of some top level officials to appreciate the need for reform," and of an "inability of those who support reform to produce effective policies."[25] Thus, the U.S. Agency for International Development reports that it has adopted a kind of wait and see approach to assistance in Turkmenistan—i.e., "to maintain a modest presence . . . watching after U.S strategic interests and looking for opportunities to support reformers, especially non-governmental organizations and the private sector."[26] In fiscal year 1999, the United States appropriated just under $18 million in assistance to Turkmenistan—one third the level requested for Kazakhstan, and half the amounts requested for Uzbekistan and Kyrgyzstan—targeted mostly for budget planning, privatization, and medical assistance. Total U.S. assistance to Turkmenistan through 1999 was only $176.3 million.[27]

These priorities make sense when aid is targeted primarily to countries deemed most receptive to reform. But dramatic changes are likely in the offing in Turkmenistan, particularly if a change of leadership becomes imminent. As U.S. policymakers, corporations, and donors balance the pursuit of investment opportunity with the promotion of human rights and other interests, more attention should also be paid to the future stability of Turkmenistan as a whole. More independent research should be supported to better understand the informal workings of Turkmenistani society, and the schisms and pressure points noted above, more deeply. Likewise, U.S. policymakers might consider expanding funding to support greater information flow in Turkmenistan, and to assessing not only what programs are conducted on the ground, but how they are implemented. Program goals and activities must be shaped to fit the informal context within Turkmenistan; and more vetting, follow-up, and monitoring of programs should be pursued, perhaps particularly among the law enforcement community where President Niyazov himself has indicated that many officials are deeply involved in criminal activity.

Finally, while U.S. reform programs are likely to be effective at best at the margins, ties should also be pursued and expanded not only with registered institutions, but with all levels of the population. It is more than likely that wide ranging change will begin to occur in Turkmenistan well before pipelines are up and running and energy revenues begin to flow. Exposing greater numbers of Turkmenistan's citizens to the outside world—through the media, expanded exchanges among the young,[28] more small partnership activities, in-country projects, and the like—may be one more way to support alternative visions among those who may ultimately be tasked with building a quite different Turkmenistan in the not too distant future.

NOTES

1. While Turkmenistan is roughly the size of California, it contains less than 15 percent of California's population, or only 4.4 million people.

2. See "Oil and Gas News," *Central Asian Monitor 1* (1999): 18.

3. For some of the public debates, see "Rival Gas Suppliers Race for Turkish Market," *Caspian Investor Newsletter*, February 1999; "Black Sea Gas Pipe Start Seen in Autumn," *Reuters*, March 24, 1999; and "Playing Pipeline Politics in Turkey," *Financial Times*, March 16, 1999.

4. The reasons are reportedly two-fold: the high cost of extraction and transportation of the hydrocarbons; and Turkmenistan's requirement for more advantageous terms, including adjustments to the tax regime for Mobil and Monument at the Garashyzlyk field and Monument at the Chelken field. See, for example, *ITAR/TASS*, March 12, 1999, citing reports by London's International Oil Exchange.

5. The government of Turkmenistan reports that 690,000 tons were gathered, against a planned target of 1.5 million.

6. On the official market, one dollar equals 5,200 manat. The unofficial dollar exchange rate has soared almost three times higher to about 13,000 manat to the dollar. See "News and Comments," *Central Asian Monitor 1*, 1999: 37–38.

7. See "USAID Congressional Presentation for FY 1999: Turkmenistan," 1999, <http://www.info.usaid.gov/pubs/cp99/eni/tm.htm> (March 12, 1999).

8. "Summary for Turkmenistan: Third Quarter, 1998," *Economist Intelligence Unit* (London, 1998).

9. Between the mid-1980s and mid-1990s, for example, infant mortality (or the number of children per thousand dying before reaching the age of one), officially dropped by almost one third in a number of categories, and from numbers already artificially low. Between 1985 and 1992, for example, the number of children dying from infectious or parasitic diseases dropped from 122 to 107 per 1,000, and the number of children dying from breathing disorders dropped from 297 to 190. Interestingly, the number dying from conditions stemming from the perinatal period rose by almost one third, from 49 to 69 per 1,000. See State Committee of Turkmenistan on Statistics, *Narodnoe khoziaistvo Turkmenistana v 1992 godu* (*The National Economy of Turkmenistan in 1992*), (Ashgabat, 1993), 29.

10. This has also been found in focus group work in Turkmenistan. See Arustan Zholdasov, JNA Associates, Inc., Tashkent and Expert Center, unpublished findings.

11. These include the contamination of soil and groundwater with agricultural chemicals and pesticides; water logging of soil due to poor irrigation methods; salinization; Caspian Sea pollution; diversion of a large share of the flow of the Amu Darya into irrigation; contaminated drinking water; desertification; and other problems.

12. On efforts to resolve the conflicts over water and other issues, see chapter 7, "Regional Cooperation in Central Asia and the South Caucasus," by Martha Brill Olcott.

13. President Niyazov continues to express concern over the burgeoning drug trade. According to official reports, of the seven hundred people sentenced in 1998 for the highest crimes in Turkmenistan, 90 percent had committed a crime connected with narcotics trafficking. As President Niyazov noted, crimes connected with drugs continue to grow each year. See Arustan Zholdasov, "President of Turkmenistan Saparmurat Niyazov announces a moratorium on the death penalty," local press review for JNA Associates, Inc. and Tashkent Expert Center, January 1, 1999.

14. The ethnic breakdown of Turkmenistan's population is reportedly: 77 percent Turkmen; 9 percent Uzbek; 6–7 percent Russian; 2 percent Kazakh; and 4–5 percent other nationalities.

15. For a useful discussion of these issues, see Michael Ochs, "Turkmenistan: The Quest for Stability and Control," in *Conflict, Cleavage, and Change in Central Asia and the Caucasus*, ed. Karen Dawisha and Bruce Parrott (Cambridge: Cambridge University Press, 1997).

16. Such as mandating a minimum number of workers who must be maintained by privatized enterprises for a specified period of time. For a useful discussion of these and other issues, see "Country Commercial Guide: FY 1999: Turkmenistan" (report prepared by the U.S. Embassy, Ashgabat, released August 1998), <http://www.state.gov/www/about_state/business/com_guides/1999/europe/turkmen99.html> (January 11, 1999).

17. Official Turkmen data suggest that investment picked up in early 1998, rising to about $180 million, or about one third of GDP, and to over 150 large-scale construction projects valued at $3.6 billion. But the quality of investment is lower, as many projects depend on import substitution, generating no exports; many are "prestige projects" with a low rate of return; and most of the investment costs are being met by the Turkmenistani government whose foreign debt continues to rise. See *Economist Intelligence Unit,* "Summary for Turkmenistan, Third Quarter, 1998" (London, 1998).

18. All disputes between U.S. investors or importers and the government of Turkmenistan have been due to the government's decision not to honor signed contracts as originally written. For a discussion of this and other issues, see "Country Commercial Guide: FY 1999: Turkmenistan."

19. At a cabinet session on March 19, 1999, President Niyazov signed a resolution for the drafting of a strategy for socioeconomic reform until the year 2010, and for establishing a state commission to implement those reforms under the president's leadership. The strategy was scheduled to be approved in December 1999.

20. See *Economist Intelligence Unit*, "Summary for Turkmenistan, Third Quarter, 1998."

21. Quoted in U.S. Department of State, *International Narcotics Control Strategy Report, 1997*, released by the Bureau for International Narcotics and Law Enforcement Affairs (Washington, D.C.: U.S. Department of State, March 1998).

22. According to Turkmenistan's Constitution, the chairman of Turkmenistan's Parliament, the *Mejlis*, would assume the presidency and then call for elections in the event of the death or permanent incapacitation of the president. The fact that the *Mejlis* itself is run by President Niyazov raises many questions about how this would work in practice.

23. See Stephen Kinzer, "Can New Caspian States Handle a Gusher of Oil Cash?" *International Herald Tribune*, January 4, 1999.

24. See Curt Tarnoff, "The Former Soviet Union and US Foreign Assistance," *Congressional Research Service Issue Brief*, updated Feb. 4, 1999, 2.

25. See *U.S. Government Assistance to and Cooperative Activities with the New Independent States of the Former Soviet Union: FY 1998 Annual Report* (prepared by the Office of the Coordinator of U.S. Assistance to the NIS, Pursuant to Section 104 of the FREEDOM Support Act, January, 1999), 56.

26. See "USAID Congressional Presentation: FY 1999: Europe and the NIS," <http://www.info.usaid.gov/pubs/cp99/eni> (March 15, 1999), 3.

27. *U.S. Government Assistance to and Cooperative Activities with the New Independent States of the Former Soviet Union: FY 1999 Annual Report*, U.S Department of State, January 2000.

28. According to USAID, in 1998 thousands of secondary school students applied for only thirty slots in USIA's Future Leaders' Exchange (FLEX) Program that brings students for homestays and high school study in the United States. See *U.S. Government Assistance to and Cooperative Activities with the New Independent States of the Former Soviet Union: FY 1998 Annual Report*, 59.

7

Regional Cooperation in Central Asia and the South Caucasus

Martha Brill Olcott

The collapse of the Soviet Union is creating new dependencies and interconnections in the Central Asian and South Caucasus regions as those living in these young states are forced to think about geography in new ways in order to be able to sustain their independence and make their states secure.[1] The best publicized of these new kinds of cooperation are energy-based and relate to Caspian Basin oil and gas reserves, but there are a host of other kinds of new relationships developing as well.

The fate of Central Asia is no longer separate from that of the South Caucasus, and the eight newly independent states of both regions are coming increasingly to recognize this. It is unclear how closely intertwined these states will become and what real benefit will accrue to each as a result of this cooperation. Some of these relationships are the products of weak states rather than strong ones. These include the growing transit of drugs, weapons, and "seditious" ideas. Many of the interconnections that have developed over the past seven years, though, result from the efforts of the leaders in these states to place old, Soviet-era ties on new, mutually beneficial footings.

This chapter shows how ten of these linkages are developing—focusing on those that are most critical to the economic and political well-being of the states of this region. Many of these relationships are directly linked to the development of oil and gas reserves, while others could help to impede this process. This chapter looks at these interconnections primarily from the point of view of the states themselves and the choices that they are making to cope with independence. Most of these new linkages are still fragile in nature, and it is unclear whether or not they will be able to flourish. If they do, the states of the region will all be strengthened. If they do not, then the states of Central Asia and the Caspian region will pose dangers to one another as well as to their neighbors both near and far.

NEW GEOGRAPHIC UNDERSTANDINGS

In the Soviet era, Central Asia and the South Caucasus were considered two largely separate entities. Kazakhstan was often not even considered a part of the "Central Asia" grouping. Since these republics became independent, however, new political and economic realities have joined their fates, and it has now become commonplace to refer to the "Caspian Region" as encompassing all eight states. The region, of course, also includes Russia and Iran, but U.S. analysts often assume that there is an inherent conflict of interests between these two states and the remaining eight newly independent nations.

Sometimes dealing with one's nearest neighbors is problematic, as such states often are traditional rivals. This has been particularly true in the South Caucasus, where the Azeris and Armenians have been at war for over a decade, and the Georgians and the Abkhaz have been clashing almost as long. Relations between the Central Asian states are much better, but there is no shortage of latent antagonisms in this region as well. Increasing the "number of chairs at the table" can sometimes have its benefits. Redefining the region as a "trans-Caspian" one for many purposes opens new opportunities for all involved. These opportunities, though, are still sharply restricted by the ongoing hostilities in Armenia and to a lesser extent by the conditions in Tajikistan. These two conflicts leave six core states currently able to cooperate.

All eight states could in theory benefit from standing together. However, it is a difficult task for new states to maximize their foreign policy potential. Each of these states must carve out a distinct state identity, and each would like to do so in a way that regulates or restricts Russia's role. The Central Asian and South Caucasus states are also uncomfortable placing too much reliance on their immediate neighbors, and neither Central Asia nor the South Caucasus are likely to develop into regions marked by strong cooperative institutions.

In a strange way, the Central Asian republics were more equal before being granted statehood than afterward. Prior to independence, Moscow was the focus of their frustration, and the central government treated these regimes in a largely undifferentiated fashion. This was especially true in the late Gorbachev years, when the four Central Asian states and Kazakhstan, as they were then commonly known, were a common target of a campaign that explicitly linked the region's Muslim heritage and traditions with its disappointing economic performance and pattern of political corruption.

Although Russia may sometimes still seem to be a common enemy, the policies Moscow follows often make sharp distinctions between the various newly independent states, and Russian policymakers frequently choose to play one against another. At the same time, as part of the process of defining statehood, each is free to decide for itself how to respond to Russia. Differing policies toward Russia are but one of a host of critical ways that the various Central Asian states try to carve out a distinct existence for themselves. Their efforts to uniquely define themselves have generally increased the atmosphere of competition that already existed between republics. Uzbeks and Tajiks had long been traditional rivals: the Tajik elite feels ousted from Bukhara and claims much of Uzbekistan as theirs by historic right. In reality though, the Uzbeks have long dominated northern Tajikistan, as well as southern Kyrgyzstan and southern Kazakhstan. The Uzbekistan-Turkmenistan border region is another point of dispute: the Turkmen claim the history of Khwarazm, one of the oldest centers of civilization in Central Asia, while the Uzbeks make the same claim. The Tajikistan-Kyrgyzstan boundary is also a source of contention.

For most of the Soviet period, such competing national claims were the cause of ill will or minor skirmishes. Now, as the fight over Nagorno-Karabakh has shown, they can be relatively easily transformed into wars between nations. The Central Asians got their first real wake-up call in 1990 when fighting broke out between Kyrgyz and Uzbeks in Kyrgyzstan's Osh region. Uzbekistan's President Islam Karimov determinedly refused to intervene, but part of the price he extracted for that was the right to regularly cast a protective glance over the Uzbeks of Kyrgyzstan in the future, as well as greater vigilance toward interethnic relations on the part of the Kyrgyz.

The war in Tajikistan, which peaked in 1992 but which has never fully abated, has created heightened concern for exacerbating interethnic tensions. The war itself is not an interethnic struggle, but rather an inter-elite one. As such it has made the governing elite in each of the Central Asian states more concerned with the need to maintain stability, for each has seen the chaos of Tajikistan as a blueprint for what could happen in its own country if a serious struggle for power were to develop.

This elite insecurity has implications for the democratization and political development of the region. Elites that fear instability may crack down on opposition activities. Opposition groups repeatedly have been backed into corners throughout the region precisely for this reason. But this tactic could easily produce the opposite of the desired effect, inflaming the opposition and making the country not only less democratic, but less stable as well.

DEVELOPING CASPIAN OIL AND GAS RESERVES

There has also been important outside pressure to redefine the South Caucasus and Central Asia as a single region. The main reason for this shift in thinking has been the rise in importance of the area's oil and gas reserves. Almost all of these countries are affected by energy development: the region's largest reserves, both proven and possible, are located in Kazakhstan; Azerbaijan and

Turkmenistan have large reserves; Uzbekistan has enough oil and gas to export to its neighbors; and Georgia has high hopes of benefiting from new transport routes that will bring the region's oil and gas to market.

Soon after independence, predictions began to spring up that the region would become a new North Sea.[2] Foreign expectations skyrocketed as projections were made that these reserves could help meet the growing energy demands of the next century while avoiding politically and economically risky reliance on current energy providers.

One of the earliest foreign investments in the post-independence Caspian was Chevron's deal with Kazakhstan, signed in 1993, that gave the U.S. oil company rights to develop Kazakhstan's huge Tengiz field. This multi-billion-dollar deal was also a symbol of how energy could become a key to securing these countries' real independence from Russia. Chevron had been negotiating with Moscow for the rights to Tengiz since 1990, but had to begin a major new negotiation effort directly with the Kazakhstani government in 1992.[3]

The history of the Tengiz field also demonstrates, however, that the independence afforded Caspian countries by selling energy development rights to foreign companies is qualified by complications in transporting the energy out of these countries. Very few countries in the world are landlocked, but every country in the Caspian region except Georgia and Russia is included in this number. The Caspian itself is an inland sea and has no outlet to a major ocean. In the case of the Tengiz, the Caspian Pipeline Consortium (CPC) project to lay a new pipeline to Russia's Black Sea port of Novorossiisk has been subjected to numerous delays and is now scheduled to receive its first shipment of oil in July 2001.[4] While there are plans to construct pipelines through China, Turkey, and Iran, for now Kazakhstan remains largely dependent on Russia when it comes to exporting its oil to the world market.

The Caspian pipeline issue has become an important geopolitical one, as the U.S. government has been very public in its support for pipelines that bypass Russia and thus shore up the independence of the states of Central Asia and the South Caucasus. The region's geopolitical importance has been most visible in the case of U.S. support for a so-called main export pipeline for Azerbaijani oil, to run from Baku to Turkey's Mediterranean port of Ceyhan, which would not pass through Russia at all.[5] Plans for a trans-Caspian pipeline would allow oil from Kazakhstan to be linked up to this pipeline as well. Western interest in the transport of Caspian oil and gas has substantially complicated the relations of all these states to Russia. In Kazakhstan in particular, it has helped shape the national identity of the state, allowing the Kazakhstani government to become much more assertively pro-Kazakh. As western interest in its oil and gas has increased, and as Russia has simultaneously weakened, the Kazakhstani government has become less interested in accommodating the cultural and political demands of the country's large Russian population, and the decision to move the nation's capital north from

Almaty to Akmola (now renamed Astana) was in large part dictated by the desire to stake out a stronger claim to the northern half of the country.

The complications in transporting the region's energy resources have also created new links among the states of the region. For these new links to work, the region must get beyond old rivalries and personal jealousies. Azerbaijan is expected to become a focal point for the region's energy distribution, because without trans-Caspian pipelines to take Turkmenistan's gas and Kazakhstan's oil to Baku for export, the proposed east-west routing will not be economically viable.

While Turkmenistan and Kazakhstan currently support these plans, the relationship between the Central Asian states and Azerbaijan is a complex and potentially unstable one. Turkmenistan and Azerbaijan have conflicting claims to the Kyapaz (also called Serdar) oil field in the Caspian Sea. But there is a more deep-seated rivalry between the leaders of these states. This rivalry is partly personal, the product of Heydar Aliyev's longer and more distinguished Soviet-era career, and partly cultural, from the Turkmen fears that Azeris will somehow best them. For now, a route through Azerbaijan seems preferable to shipping through Russia, but Baku may prove to have less to offer as a long-term partner than does Moscow.

Pipeline politics is also affecting the internal dynamics of some of the Caspian states. For Georgia, pipelines passing through its territory might be its only hope of economic survival. The Baku-Supsa pipeline, an option for shipping oil from Azerbaijan that is currently transporting modest amounts of crude, traverses many miles of Georgian territory and could provide it with substantial tariff revenues if it were enlarged to become a main export route.[6] Georgia must pay for the possibility of this advantage, however, by dealing with outside countries that have vested interests in the outcome of pipeline routing. The Baku-Supsa pipeline is an alternative to Baku-Ceyhan as an "east-west" route that would not pass through Russia. The capacity of this route is much more limited than Baku-Ceyhan, but its existence does decrease the region's potential dependence on Russia. This gives the Russians an obvious reason to undermine it by continuing to aggravate the separatist crises in Ossetia and especially in Abkhazia. Russian actions in Georgia are hard to interpret, and a host of different groups seem to have their own reasons for wanting a weak Georgian state, but flare-ups in these conflicts could easily sway investors, who are already concerned about the security of pipelines in such a volatile part of the world. This concern has made Georgian President Eduard Shevardnadze into a greater political accommodator than might otherwise be the case, both to try to defuse his country's many ethnic conflicts and to show foreign investors that he is an ardent supporter of democratic reforms.

The need to transport oil and gas is also affecting relations between the Central Asian states and a number of their neighbors outside the former

Soviet Union. Three Central Asian states border on Afghanistan, and they have had very different responses to that country's civil war. Turkmenistan, which hopes to export gas to Pakistan and India via Afghanistan, has been the Central Asian state most eager to reach accommodation with the Taliban. By contrast, Uzbekistan and Tajikistan have felt directly threatened by the prospect of a Taliban victory, in large part because the ethnic Tajiks and Uzbeks native to Afghanistan would then be on the losing side.[7]

Ethnic issues have also helped shape relations toward Iran, and they have affected the receptivity of the various oil and gas producing states to a southern pipeline route. Turkmenistan and Azerbaijan both border on Iran and have had very different attitudes toward close cooperation with the regime in Tehran. Some arrangements between Iran and the Caspian countries have already been signed. Turkmenistan, for example, signed a deal in May 1997 that would clear the way for its gas to be transported through Iran to Turkey starting in the beginning of the next century.[8] Kazakhstan also is currently exporting oil through Iran. The Iran option is severely limited, however, by U.S. restrictions on trade with that country. While Turkmenistan and Kazakhstan are sensitive to U.S. pressure, they are not fearful of doing business with Iran; they simply want the best deal possible. For national reasons, the Azeris have a much more tenuous relationship with the Iranians. The Azeris view themselves as a divided people and hold vague dreams of someday reunifying their country with "southern Azerbaijan," which is under Iranian rule.

The desire to appease a powerful neighbor seems to have played an important role in Kazakhstan's decision to turn down western oil companies' bids and award China the rights to develop the huge Uzen oil field, located approximately 3,000 kilometers from the Chinese border. In turn, China promised to build a $9.6 billion pipeline over eight years. This was a sign of China's growing prominence in Kazakhstan's energy sector, as the Uzen field is second in size only to the Tengiz.[9] Though China National Petroleum Corporation (CNPC) is currently involved in a dispute with the Kazakh government over the layoff of 2,000 workers, it has no intentions of leaving the Kazakh market. However, the project slowed down as CNPC contributed just around 60 percent of the planned investment in 1999.[10] The China-Kazakhstan connection was reaffirmed in early 1999 when Presidents Nursultan Nazarbaev of Kazakhstan and Saparmurat Niyazov of Turkmenistan emerged from an April 1999 meeting saying they would give priority to energy export lines through China in the next century.[11]

While the Caspian region is now assumed to contain 2.7 percent of the world's proven oil reserves, South America contains 8 percent of the world's reserves and the Middle East holds 55 percent.[12] Thus, while energy will be important to the future of the region, we should not overemphasize the interstate dependencies created by the pipeline issue. Energy links ultimately may be only one path in a maze of different issues and ties across the region.

NEW TRANSPORT LINKAGES

A goal of all eight states is to expand the capacity for trade within the Caspian Basin and between the region and the rest of the world. One major effort to meet this need is the Transport Corridor Europe-Caucasus-Central Asia (TRACECA) program. This program, started in 1993, is a cooperative effort between the European Union (EU) and the states of the South Caucasus and Central Asia.[13] The group has as its goal the recreation of the "Ancient Silk Road," a transport corridor that would facilitate trade between Europe and the Caspian region. The key point about this trade route is that it would not pass through Russia, but would run instead from Europe through the Black Sea, the South Caucasus, and the Caspian Sea to Central Asia. Indeed, one of the EU's stated goals for TRACECA is to strengthen the independence of the Caspian states. TRACECA has worked with the EU and international lending institutions like the European Bank for Reconstruction and Development and the World Bank to obtain technical assistance for highway construction, port rebuilding, railway modernization, and other projects to shore up the transportation infrastructure in the region.[14]

The TRACECA program has proved to be an important forum for regional cooperation. Five working group meetings have been held since its inception, and a major TRACECA conference was held in Baku in September 1998. The Baku conference adopted a "Basic Agreement" that called for ensuring these countries' access to the world market by road, rail, and commercial navigation. It also encouraged cooperation on traffic security, environmental protection, legal transport regulations, and other related goals.[15] A symbolically significant aspect of the conference was Azerbaijani President Heydar Aliyev's invitation to Armenia to send a delegation despite the unresolved conflict in Nagorno-Karabakh.

TRACECA is important both as a facilitator of cooperation among the Caspian states and as a bridge between the Caspian and the rest of the world that does not depend on Russia. There are also plans to greatly expand trade to Asia through China. The major rail link across the region (from Druzhba in Kazakhstan to Urumchi in China) was opened in 1992, and new links through China, across Central Asia and Iran, and on to Turkey and Europe promise to cut several days off the transport time for old Soviet routes. The old routes will probably not be abandoned, however. The amount of freight that needs to be shipped will support two routes, and contacts and familiarity with old Soviet ports will continue to make them attractive. Highway connections through Central Asia into China are complicated by the harsh climate and high mountains on the Kyrgyzstani-Chinese border. The Karakoram Highway from Urumchi to Pakistan is currently the main highway link between Central Asia and China, and in 1996 Kazakhstan, Kyrgyzstan, Pakistan, and China launched an improvement project that promised to make the road passable year round. Additionally, TRACECA is helping to fund upgrades to a road from Kyrgyzstan's Osh region that will meet up with a road the Chinese are building to the Kyrgyzstani-Chinese border at Erkecham. A new customs point opened on that road in July 1997. There are several links to the Karakoram Highway from Bishkek, one of which is also the shortest route from Almaty, Kazakhstan, to China. Despite complications, such as lack of freight storage facilities in some places, the Karakoram Highway remains the only feasible way to move freight from Central Asia to Indian Ocean ports.

Since these alternative transit routes are largely in the planning stage, it is still too soon to predict how well these states will cooperate with each other to make them a success. Preliminary evidence, though, is not encouraging. Border crossings within Central Asia remain generally slow, and the Turkmenistan-Uzbekistan border is a real bottleneck on the highway into the region from Iran (and Turkey). Efforts to achieve cooperation on the railroad system have also been very slow to materialize, and there is still no unified tariff regime for the region.

While Russia is becoming an increasingly less important trading partner for the Caspian states, it still plays a critical role in these states' economic well being. Despite Caspian leaders' insistence that their countries would not be adversely affected by Russia's ruble devaluation in August 1998, Russia's financial woes had clear repercussions for them. Most notably, the sharp drop in trade with Russia led to major industrial stoppages in Kazakhstan and Georgia, and Azerbaijan and Armenia were hurt by the sudden cutback in remittances from citizens working in Russia.[16] The region's currencies have been affected by the crisis as well. Kyrgyzstan's som, once considered the region's most stable currency, fell from 17 to 30 to the U.S. dollar between July and November 1998.[17] Kazakhstan's tenge remained stable immediately following the crisis, but it fulfilled analysts' fears when it began to drop in early 1999. The Kazakhstani government responded with a decision to float its currency on April 4. Prime Minister Nurlan Balgimbaev explicitly stated that Kazakhstan had been pushed into that action by the reduction in the competitiveness of its exports that largely resulted from the drop in Russia's currency.[18] Thus, despite initiatives like TRACECA, which attempt to extricate the Caspian from its reliance on Russia, that country continues to hold a disproportionate amount of economic sway over the region.

THE NARCOTICS TRADE

The porousness of Central Asia's borders has caused undesirable types of trade to expand as well, including most prominently narcotics trafficking. When the drug trade takes hold, it is very difficult for healthy economic forces to assert themselves. Narco-business in Afghanistan figures prominently in the Caspian-region drug trade and poses a direct threat to economic development in Tajikistan and to a lesser extent in Kyrgyzstan and Turkmenistan. Transit trade also puts Uzbekistan and Kazakhstan somewhat at risk, and the easy ability to trade drugs for cash makes any nascent opposition group potentially self-financing.

Drug dealers in neighboring states have found that the Caspian countries can provide markets for their drugs and smugglers for their operations. In Uzbekistan alone, sixteen tons of narcotics have been destroyed since 1994, and security forces confiscated 23 tons between 1991 and 1996.[19] The most common drug to pass through the region is opium, and a substantial heroin trade is also developing. A drug bust in Tajikistan in August 1997 recovered 270 kilograms of raw opium that Afghanistani smugglers were trying to bring into the country.[20] Kyrgyzstan has also been a link in the Afghan drug trade. Police in Kyrgyzstan's Osh region had some success in lessening this trade in May 1998, when the arrest of two opium traffickers and the seizure of 8.4 kilograms of opium in their possession led to the closure of the so-called "Siberian channel," through which drugs had previously been smuggled.[21]

Kazakhstan's narcotics problem brings China into the picture as a player in the Caspian region's drug scene as well, as the two countries share a fairly long border. Officials confiscated 30 kilograms of raw opium from the trunk of a car in East Kazakhstan oblast' in February 1998. In a more sensational development that year, the Kazakhstan National Security Committee seized 6,000 packets of a highly addictive LSD-type amphetamine called phenchloramine hydrochloride that was produced in China. Officials said the dangerous substance was easy to smuggle, as packets could be hidden in clothes and shoes.[22]

Drugs seep into the South Caucasus states through Iran, which shares a border with Armenia and Azerbaijan. Russian troops stationed in Armenia discovered a drug channel on the Armenia-Iran border in 1997.[23] In August 1998, Georgian police confiscated 7,400 grams of opium that they said had been grown in Pakistan and shipped through Iran and Azerbaijan to Georgia.[24] Azerbaijan reported that in 1998 more Iranians than citizens of any other country were arrested on drug charges.[25]

Security forces in the Caspian stage hundreds of the types of drug busts described above, but these types of operations are not sufficient to halt the region's narcotics trade. Border controls are far from impregnable, and corruption in the governments of the region sometimes counteract the efforts of anti-drug police. In some cases, these police themselves, who are often paid minimal salaries, are lured by the promise of money into helping the very smugglers whom they are charged with eliminating. In March 1998, for example, Kyrgyzstan began criminal proceedings against a group of police officers accused of operating an elaborate drug-smuggling network. The operation had allegedly penetrated the highest levels of power in the Osh region, as one of the defendants, Adyl Madazimov, was the former head of the anti–drug-trafficking section of the regional Internal Affairs Department.[26]

Combating the drug problem could provide an important impetus for regional cooperation if the Caspian states were to organize their efforts. So far, however, most anti-drug actions have been taken unilaterally. Whether or not they decide to cooperate in solving the problem, though, the region's leaders have no choice but to acknowledge the links the drug dealers have formed among their countries. Ethnic issues have also made it harder to combat the drug problem, for dealers are able to take advantage of extended families and various forms of kin relationships to form ties that cross state boundaries.

MIGRATION

Another way in which the countries of the Caspian region have become intertwined since independence is through the increase in interstate migration. There are generally two reasons for this increase: the rise in violent conflict in the region and the perceived isolation of ethnic diaspora in now-independent states. Both of these types of migration are helping to make each Caspian state more mono-national.

The first kind of migration is typified by the refugee situations in Azerbaijan and Armenia, which have been locked in a dispute over Nagorno-Karabakh since the collapse of the Soviet Union. At the end of 1997, 218,000 of Armenia's 219,150 refugees and asylum seekers originated in Azerbaijan. In Azerbaijan, 188,000 of its 244,100 refugees and asylum seekers came from Armenia,[27] and more than twice that number poured out of parts of Azerbaijan that Armenia had captured in the conflict.[28] The civil war in Tajikistan has forced large-scale migration as well. Ethnic Kyrgyz who fled Tajikistan into neighboring Kyrgyzstan make up the majority of Kyrgyzstan's 14,500 registered refugees. Kyrgyzstan has also received ethnic Tajiks fleeing the war, but they are being slowly repatriated.[29] Russians fled from Tajikistan's conflict as well, as 41 percent of Tajikistan's Russian population left the country between 1989 and 1995.[30]

The group that best exemplifies the second reason for migration is the huge Russian diaspora. Soviet policies encouraged Russians to move into other constituent republics of the Soviet Union, and, as a result, thousands of Russians found themselves in foreign lands when the Union collapsed. Secularized Russians whose careers had been connected with Moscow, as well as Russians who felt threatened by the nationalism of the post-independence elites, responded to their new status by leaving their newly independent countries.

In Kazakhstan, over a third of the population at independence was ethnically Russian, and many Russians there felt cut off from their homeland. Language laws, while conciliatory in comparison with those in former republics like the Baltics, added to Russians' fears that they would become second-class citizens if they stayed in Kazakhstan.[31] As a result, between 1992 and 1996 the net out-migration of ethnic Russians was over 700,000 people, and over the past ten years Kazakhstan has seen its Russian population decline by 18 percent.[32] A 1994 survey of Russians and Kazakhs, 71 percent of whom planned to leave the country, asked respondents about their motivations for emigrating. Of Russians surveyed, 59.4 percent gave their reason as "there is no future for my children here"; 40.2 percent cited "material difficulties"; and 28.2 percent said "not knowing the [Kazakh] language created work-related difficulties."[33] The desire to leave Kazakhstan remains fairly high. A 1998 study of ethnic Russian university students in Almaty and Astana found that only 33 percent in Almaty and 39 percent in Astana planned to stay in Kazakhstan after their studies.[34]

Early post-independence emigration from Uzbekistan caused that country to lose an even greater percentage of its Russian population. Between 1990 and 1997, 516,000 ethnic Russians left the country, amounting to 44.5 percent of Uzbekistan's total emigration. Total net outmigration from Uzbekistan to Russia during that period was 451,400 people.[35] In the 1994 study cited earlier, 38.9 percent of Russians wanting to leave Uzbekistan pointed to problems with the Uzbek language as their motivation; 19.4 percent felt they were being forced to leave.[36]

Another migration issue involves people from the Caspian states who work in Russia. Many citizens of states bordering on Russia have sought better employment there and regularly send remittances back to their families. One of the most striking examples of this is in Azerbaijan, where an estimated one in three working-age men is employed in Russia.[37]

The trend of increased migration around the region is slowly altering the ethnic balance of the Caspian states. In general, the trend is toward ethnic consolidation, as the titular nationality is everywhere increasing its demographic domination. In Kazakhstan, for example, ethnic Kazakhs comprised a majority of the population in 1997, after years of being the only titular nationality not to hold a majority in a post-Soviet country.[38] As the states become less multi-ethnic, their populations may come to be less tolerant of the numerous minorities that still live among them. Migration issues will also continue to create new dependencies and connections among the states of the region, as countries turn to diaspora communities and kindred states to help fill the gaps left by the outmigration of their indigenous European populations.

EDUCATION

New trends in education have affected the links among countries in the Caspian region as well. Increasing nationalism in these states and their desires to distinguish themselves from Russia have been reflected in the types of education they offer. Since independence, there has been a trend away from Russian language education, and each country in the region has seen a steady decline in the number of its young people who go to Russia for university studies. (See Table 7.1.)

A number of alternatives have sprung up in place of Russian-oriented education. Many students have begun studying in Turkey and the Middle East, emphasizing the Caspian countries' cultural ties with eastern states. Religious education is also becoming more prevalent throughout the region, and this is a development that is oftentimes not viewed very favorably by the secular-oriented governments. While none of these regimes wants to sever ties with the outside Muslim world, all wish that the infiltration of religious ideas could be better controlled. This is especially true in Uzbekistan, where the Ministry of Higher and Secondary Specialized Education and the Institute of Oriental Studies recently announced that they will invite the nation's leading educators to a seminar on teaching basic religious principles.[39] However, there is still no sufficient evidence of significant cooperation on education among Caspian states themselves.

An elite sector has also been singled out for education in the western tradition, as these states attempt to train a new generation of leaders to function in the international arena. Scholarships have been established for youth to study in the West. In 1998, Kazakhstan opened the first of six planned government-funded Kazakhstani-American schools. But these trends in education have negative as well as positive consequences for society. The proliferation of opportunities to study abroad for students in wealthy and elite families is widening the gap between the elite and the masses. In the same year that Kazakhstan opened the first of its Kazakhstani-American schools, for example, it came to light that the country had previously been forced to close over 100

Table 7.1: Students from the Caspian States Studying in Russian Universities
(by academic year)

	1992–93	1993–94	1994–95	1995–96	1996–97	1997–98
Armenia	5,600	3,600	2,200	1,700	1,200	1,100
Azerbaijan	5,800	4,000	2,600	1,700	1,100	1,000
Georgia	11,400	7,500	4,900	3,100	1,900	1,400
Kazakhstan	22,000	19,000	14,300	11,600	11,100	10,700
Kyrgyzstan	4,800	3,500	2,500	1,700	1,300	1,200
Tajikistan	2,600	2,100	1,300	700	500	400
Turkmenistan	1,900	1,600	1,000	700	500	400
Uzbekistan	8,100	6,000	4,500	3,200	3,200	3,300

Source: GOSKOMSTAT, _Rossiiskii Staticheskii Ezhegodnik_, 1998 (Moscow: GOSKOMSTAT, 1998).

public schools because of a supposed lack of funds.[40] The previous year, the Red Cross had estimated that 11 percent of Kazakhstan's children did not attend school because they lacked adequate clothing.[41] This growing rift between the elite, who reap the benefits of government-funded education, and the masses, who are often left out of these types of government programs, is echoed in many societies in the region and will continue to have domestic repercussions. These tensions could well be keenly felt when the current generation of political leaders prepares to pass from the scene and tries to transfer authority to successors of their choosing.

BOUNDARIES: THREATS TO SECURITY

The complex issues intertwining the fates of the Caspian states are complicated even further by the fact that most of these states lack fully demarcated international borders. Boundary demarcation has been postponed in many of these countries because it is expected to be a very sensitive and controversial process. In this part of the world, ethnic groups straddle administrative boundaries, and national diaspora are scattered around the region.

The situation in the South Caucasus is far more volatile than that in Central Asia. There are separatist movements in Georgia's Abkhazia region and Azerbaijan's Nagorno-Karabakh. During the Soviet period, both were autonomous regions within their respective Soviet republics, and both have demanded independence. Violent ethnic clashes took place in Abkhazia in 1989, and tensions were intensified when nationalist Zviad Gamsakhurdia became president of a newly independent Georgia in 1991. Fighting continued, and Abkhazia gained de facto independence in 1993. Ethnic tensions remain, and Russian-sponsored peace negotiations have failed to solve the problem. There is also the question of the status of southern Ossetia (in Georgia), and its relationship to northern Ossetia (in Russia).

The state-building process in both Armenia and Azerbaijan has been shaped by the dispute over Karabakh. The majority of the population of Azerbaijan's Nagorno-Karabakh region is ethnically Armenian, and the region declared its independence from Azerbaijan within the Soviet Union in 1989. The conflict continued after independence, and by 1993 the Armenians had won control of almost one-fifth of the territory within Azerbaijan's administrative boundaries. The Minsk Group of the Organization for Security and Cooperation in Europe (OSCE) has tried to facilitate negotiations between Armenia and Azerbaijan, but each plan the Group puts forward has been unacceptable to one or the other of the combatants.

There are also some latent disputes in the region. The Lezgin ethnic group, for example, straddles the border between Russia and Azerbaijan, with the majority of Lezgins living in Russia's Dagestan. There have already been incidents of disturbance among these people. Members of Georgia's Armenian population have also become more vocal in recent years.

By contrast, while the borders in Central Asia remain unclear, it is significant that the situation along most of the disputed borders has remained largely peaceful. These issues have caused enough uneasiness, however, to make Central Asian officials postpone final border demarcation. The region's best publicized potential border dispute is between Russia and Kazakhstan, where the government of the latter has feared the rise of a Moscow-supported separatist movement from the time of the state's creation in 1991. This is one of the major reasons why Kazakhstan's government decided to move its capital from Almaty in the southeast to Astana in the center of the country.

Russians still make up a significant percentage of Kazakhstan's population despite substantial outmigration, and they still live in ethnically consolidated groups near the Russia-Kazakhstan border. Those Russians living in Kazakhstan see Moscow as their protector and would view a definitive border demarcation as a sign of their final abandonment by Russia.

The one state that everyone fears is Uzbekistan, which has borders with all four of its Central Asian neighbors and has a diaspora population living in each as well. Uzbekistan's president Islam Karimov made neighboring states uneasy when he chose the historic figure of Timur (or Tamurlane, as he is known in the West) as the symbol of Uzbekistani statehood. Timur was a fourteenth century leader who commanded an empire encompassing most of the Caspian region.

Long-standing tensions in Kyrgyzstan-Uzbekistan relations have been aggravated recently, as Uzbeks have begun to plant private plots that creep ominously into Kyrgyzstan's Osh region. Osh was the site of interethnic violence in 1990, when the region's Uzbek population rioted to protest the denial of the broad autonomy it had demanded.

Turkmenistan's borders with Uzbekistan and Kazakhstan remain fluid as well. The stakes are high in this game because of the resources that lie in the sands under the border region. To date, these borders have not been the subject of high-profile disagreement, as has the previously mentioned dispute between Turkmenistan and Azerbaijan over demarcation of their respective sectors of the Caspian.[42]

The border issues in the Caspian region, violent and nonviolent, remain abundant. Until they are resolved, they will be barriers to normal relations between states in Central Asia and the South Caucasus.

COOPERATION IN WATER MANAGEMENT

The management of water resources is an especially acute problem in the Caspian region, particularly in the arid Central Asian lands. Water sources are unevenly distributed throughout the countries of the region, and there is no overarching structure regulating their use for irrigation and power needs. Soviet planners decided how the region's water would be distributed among its republics, but since independence, divergent national interests have often obscured the goal of regional cooperation.

Water sources in the region are concentrated in Tajikistan and Kyrgyzstan, and then flow to the heavy user states of Turkmenistan and Uzbekistan.[43] As the situation stands, water-rich countries must pay for the maintenance and operation of water management facilities, such as irrigation centers, on their territories. The Soviet government built several such facilities, including Kyrgyzstan's giant Toktogul Dam, in the 1960s, 1970s, and 1980s.[44] The artificial seas and pumping stations built during this time in Kyrgyzstan alone led to the flooding of 47,000 hectares of arable land.[45] These countries now feel that they are losing the revenue they could have gained from farming this land, as well as the money they must spend on upkeep of the facilities.

Kyrgyzstan and Tajikistan would favor an overarching management structure to regulate not only water resources, but also oil, gas, and electricity. Since independence, there have been numerous attempts to put together a regime that would deal at least with water management issues. Meeting in Nukus, Kazakhstan, in 1994, Central Asian leaders agreed to set up a regional water management system. As of 1997, there was a theoretical structure in place consisting of state-level water ministries, a Basin Water Management Organization, and an overall regional water management council.[46] By the following year, the International Fund for Saving the Aral Sea and the Inter-State Council of the Central Asian Economic Association had jumped on the water management bandwagon and were considered the leading organizations in Central Asian water resource management.[47] In recent years, regional leaders have agreed at successive summits and meetings to create various other types of regional water management organizations.

To a large extent, these regional systems have remained hypothetical and theoretical. The real business of water management has been done through a series of bilateral and sometimes trilateral arrangements, which usually consist largely of barter. In one such agreement, signed in 1997, Kyrgyzstan agreed to supply water to Uzbekistan and Kazakhstan in exchange for deliveries of gas from the former and coal from the latter. These agreements often go unfulfilled and lead to resentment and conflict. The 1997 trilateral agreement, for example, was endangered in late December of that year when Kyrgyzstan threatened to cut off water supplies to Kazakhstan, which had failed to fulfill its part of the bargain. Officials from the two countries met a few days later and made new promises to deliver their goods. Despite this "resolution" of the dispute, the three signatories to the 1997 agreement found it necessary to meet again in March 1998 to sign another very similar document. In other words, these deals provide little stability or predictability for Central Asian water management, as they must be constantly renegotiated.

A more macro-level ecological issue arising from the water situation in Central Asia involves the fate of the Aral Sea. This inland sea, which is dissected by the Kazakhstan-Uzbekistan border, is fed by the two major rivers in the region, the Amu Darya and the Syr Darya. The Amu Darya passes through the northern part of Turkmenistan before flowing through Uzbekistan to the Aral; the Syr Darya passes through Kyrgyzstan, Uzbekistan, and Kazakhstan before reaching the sea. Both Kazakhstan and Uzbekistan withdraw great volumes of water from the rivers, and Soviet planners diverted water from the rivers to various irrigation canals in an attempt to increase the region's cotton production.[48] As a result, the level of the Aral has gone down 15 meters since the 1960s, and its surface area has decreased by half.[49] The sea is now in danger of dying, and the Aral Sea basin has received international attention as an area in extreme ecological distress.

These and other ecological issues present another challenge on which the countries of the region might work together. The high transaction costs associated with the current system of water management demonstrate the consequences of these states' failure to cooperate, and perhaps in the future will serve as an impetus for increased regional cooperation.

SECURITY COOPERATION

Security links among the Caspian countries are very underdeveloped. The only formal military integration effort in the region is the Central Asian Battalion (CENTRASBAT), which involves only Kazakhstan, Uzbekistan, and Kyrgyzstan. The presidents of these countries agreed to form the battalion on December 15, 1995, at a meeting in Jambyl, Kazakhstan. The battalion was to operate under the auspices of the United Nations, with the purpose of preventing conflicts in volatile areas of Central Asia.[50] Over the past two years, the group has become a vehicle for cooperation between Central Asia and western countries, as it has been the focus of two major military exercises sponsored by NATO's Partnership for Peace (PFP) program. The first operation, called "Central Asian Battalion-97," took place in Kazakhstan and Uzbekistan and drew Georgian troops into cooperation with the three CENTRASBAT states; troops from the United States, Russia, Turkey, and Latvia also participated.[51] The second operation took place at the same time the following year, suggesting the exercises were becoming annual occurrences. "Central Asian Battalion-98" managed to pull both Georgian and Azerbaijani troops into cooperation with CENTRASBAT, U.S., and Turkish personnel.[52]

CENTRASBAT is important to the extent that it establishes patterns of military cooperation among Caspian countries. Its cooperation with western states is important as well, as it is a step, however small, toward consolidating these countries' independence from Russia's sphere of influence. Still, the major security issues for the Caspian states lie very close to home, and distrust of regional neighbors often prevents the establishment of useful integrative security systems. The civil war in Tajikistan, for example, makes other states in the region reluctant to include it in regional security structures. The South Caucasus states are even less likely to develop regional security structures, as Azerbaijan and Armenia have yet to even establish peaceable relations with each other. Certainly regional military cooperation is an unthinkable concept when two out of the three countries in the South Caucasus consider each other major security threats.

As a result of the lack of substantial cooperative security structures in the region, the militaries there are growing apart as they develop distinct personalities and modes of operation. Different languages of command presumably will become barriers to cooperation, as national languages increase their prevalence in these societies. Each state has undertaken to make the language of its titular nationality an official language, if not the only official language, and it seems only a matter of time before the once-unifying Russian language ceases to be dominant. In the absence of a mechanism facilitating regular military cooperation and consultation, it is logical to assume that command structures and military doctrines in the Caspian states also will become increasingly divergent as each country develops its own strategies apart from any institutionalized regional framework.

The danger stemming from a lack of regional military cooperation is twofold. First, if these states pursue relatively isolated military development, their mutual distrust will not be alleviated and might even be exacerbated, which would increase the risk of conflict in the region. Second, these states are militarily weak, and in most cases it is unlikely that any one of them will be able to provide for its own security needs.[53] This raises the further possibility that domestic conflicts will spill over porous borders into neighboring states. Events of August–October 1999 show how Uzbek opposition groups that were pushed out of Tajikistan as part of the reconciliation process can destabilize the situation in neighboring Kyrgyzstan. The less these states rely on each other for support in this capacity, the more they may be forced in the future to turn to outside countries—most likely Russia but down the road possibly China or even Iran—for their military needs.

INSTITUTIONAL COOPERATION

The many linkages and forms of interaction discussed here make it natural to expect that the states of the Caspian would consider developing more formal institutionally based kinds of regional integration.

All the countries of the region are members of the Russian-dominated Commonwealth of Independent States (CIS), which was conceived as a successor to the defunct Soviet Union but which has been disappointing in its effectiveness. Georgia, Ukraine, Azerbaijan, and Moldova set up a multilateral organization (originally called GUAM from the first letters of the names of the members, and after the addition of Uzbekistan, GUUAM) but it involves countries outside the region. The only organization that attempts to integrate countries solely within the Caspian region is the Central Asian Union (CAU). This first step toward Central Asian integration developed out of a 1993 summit of the five Central Asian leaders. They signed a communiqué setting up embassies in the region, agreeing to pursue the creation of a common market, and providing for the exchange of representatives to coordinate joint activities.[54] Kazakhstan, Kyrgyzstan, and Uzbekistan took the lead in putting these intentions into effect. In January 1994, the leaders of these three states signed an agreement to create a "common economic space." This document became the central and most important document to the Central Asian Union. The three states followed up on that document in July 1994 with agreements to create an interstate council of presidents and prime ministers, a permanent executive committee, a Central Asian Bank for Cooperation and Development, and councils of foreign and defense ministers.[55]

The other two Central Asian states have been absent from the CAU until recently. The original three states were wary of linking their fates with unstable Tajikistan. That country was made an observer in 1995, but only recently has it signed the central economic integration agreement. Turkmenistan has refused to join the union, sticking by its determined declared status of neutrality. It has been generally disinterested in Central Asian regional affairs.

Much like the TRACECA group, the CAU has been useful as a facilitator for regional forums but ineffective in accomplishing its stated goal of economic cooperation. The Central Asian Bank is underfunded and has not sponsored any major economic projects. The free economic zone promised by the central CAU document simply does not exist. Political tensions between Tajikistan and Uzbekistan have caused the virtual closure of their mutual border for the past two years.

Even among the three original CAU states, which enjoy comparatively good relations, differing economic conditions and strategies have hampered free interstate trade. Uzbekistan retains some price controls, and its currency is not fully convertible. Kazakhstan and Kyrgyzstan have repeatedly placed tariffs on each other's goods because differences in standards of living cause prices to vary widely between the two states.[56] The members of the CAU do not even offer preferential price-setting treatment to each other.

On the other hand, the three core CAU members do maintain good political relationships with each other, despite their lack of economic cooperation. The presidents of the three countries signed a treaty of "eternal friendship" in January 1997 at one of their frequent summits.[57] Their good relations depend to a large extent, however, on the personal relationships among their leaders.

President Nursultan Nazarbaev of Kazakhstan and President Askar Akaev of Kyrgyzstan, for example, became in-laws in July 1998 when Nazarbaev's daughter married Akaev's son. The politics of personal relationships is vital to relations among all the Central Asian states, whose leaders worked together as members of the Soviet elite. This creates an atmosphere of unpredictability in the region and a sense that everything could change when the next generation of leaders comes to power.

The future of the Caspian remains ambiguous. Energy resources may or may not prove to be great sources of wealth for these countries; but either way, these states will not see the instant prosperity they once imagined. Trade links have been established among these states and with new trading partners in the "far abroad," but Russia still looms large as an influential economic and political force in the region. Narco-business remains a daunting problem for these countries as does their inability to coordinate regional ecological, security, and economic policies. Lingering mistrust among the countries makes such initiatives difficult to carry out, and it continues to hamper progress toward demarcating borders.

These largely negative perceptions may be the result of inflated expectations about the region following the Soviet collapse. There are some positive preliminary signs that these countries are learning to rely on each other to some extent. For example, despite the fact that TRACECA and the Central Asian Union have not fully accomplished their stated objectives, they have facilitated frequent contact among the region's leaders. And while meaningful regional cooperation in solving problems like drugs and water management seems a faraway goal, the very existence of such problems creates an urgent imperative for these states to work together in the future.

NOTES

1. Central Asia is considered to encompass Kazakhstan, Kyrgyzstan, Tajikistan, Turkmenistan, and Uzbekistan. The South Caucasus includes Armenia, Azerbaijan, and Georgia.

2. The North Sea has become an important source of oil, as output rose from roughly 2 million barrels a day in 1980 to 6.1 million barrels a day in 1998, about 8 percent of world supply. Data provided by *Petroleum Market Intelligence,* published by Energy Intelligence Group.

3. For more on Tengizchevroil and its impact on Kazakhstani policies, see chapter 5, "Kazakhstan: The Long-Term Costs of Short-Term Gains," by Pauline Jones Luong.

4. The CPC includes Russia with a 24 percent interest, Kazakhstan with 19 percent, Oman with 7 percent, Chevron with 15 percent, Mobil Oil with 7.5 percent, Oryx with 1.75 percent, Russian-American Lukarco (a partnership of LukOil and ARCO) with 12.5 percent, Russian-British Rosneft-Shell Caspian Ventures with 7.5

percent, Agip with 2 percent, British Gas with 2 percent, and Kazakoil-Amoco with 1.75 percent. *Interfax Petroleum Report,* March 27–April 3, 1998.

5. Another option for the main export pipeline from Azerbaijan is a route from Baku to Supsa on Georgia's Black Sea coast. The oil would then have to be taken by tankers through the Bosporus Strait. The United States and Turkey oppose this option because of the ecological implications of increased traffic in the strait.

6. The Baku-Supsa pipeline was officially commissioned in April 1999. In its current form it cannot be considered a main export line.

7. For additional details on Central Asian views of the Afghan conflict, see chapter 10, "The Afghan Civil War: Implications for Central Asian Stability," by Shireen Hunter.

8. Martha Brill Olcott and Amy Myers Jaffe, "The Maturing of the Caspian," in *The Euro-Asian World: A Period of Transition,* ed. Yelena Kalyuzhnova and Dov Lynch (Macmillan, forthcoming).

9. *ITAR-TASS,* September 24, 1997, in *FBIS Daily Report,* SOV-97-267, September 24, 1997. Kazakhstan was able to justify giving China the rights to Uzen because China was the only country that offered to build a pipeline. Even though Uzen is second in size to the Kazakhstan's Tengiz field, it is still small by world standards and would not normally justify such construction.

10. For a detailed discussion of the situation, see *Interfax Oil and Gas Report,* vol. N04 (January 28–February 3, 2000), 419.

11. *RFE/RL Newsline,* April 9, 1999.

12. Estimates on the proven reserves of the Caspian have been published in a variety of studies, including International Energy Agency; Center for Strategic and International Studies; Woodmac MacKensie; and Manik Talwani and Andrei Belopolsky, "Geology and Petroleum Potential of the Caspian Sea Region" (working paper, Baker Institute for Public Policy, Rice University, April 1998).

13. Since the program's inception, Mongolia, Ukraine, and Moldova have joined as well.

14. TRACECA has received $200 million in investments from the EBRD and $40 million from the World Bank, as well as $50 million from the Asian Development Bank. *Jamestown Monitor* 4, no. 165 (September 11, 1998).

15. "What is TRACECA?" <http://www.traceca.org/tracecafr.htm> (July 15, 1999).

16. Many Azeris and Armenians work in Russia and send their earnings back to their families. This phenomenon will be discussed in greater detail later in the chapter.

17. *RFE/RL Newsline* 3, no. 67 (April 7, 1999).

18. *RFE/RL Newsline* 3, no. 66 (April 6, 1999).

19. *ITAR-TASS,* June 12, 1998, in *FBIS Daily Report,* TDD-98–163, June 18, 1998.

20. *ITAR-TASS,* August 6, 1997, in *FBIS Daily Report,* TDD-97–218, August 7, 1997.

21. *Vecherniy Bishkek,* May 27, 1998, in *FBIS Daily Report,* TDD-98-160, June 11, 1998.

22. Adil Urmanov, *Almaty Delovaya Nedelya,* May 22, 1998, in *FBIS Daily Report,* TDD-98-162.

23. "Segodnya" newscast, NTV, November 20, 1997, in *FBIS Daily Report,* TDD-97-324, November 20, 1997.

24. *ITAR-TASS,* August 13, 1998, in *FBIS Daily Report,* TDD-98–226, August 14, 1998.

25. Turan, January 21, 1999, in *FBIS Daily Report,* TDD-99-025, January 25, 1999.

26. *Vecherniy Bishkek,* March 19, 1998, in *FBIS Daily Report,* TDD-98–082, March 25, 1998.

27. U.S. Committee for Refugees, *World Refugee Survey, 1998* (Washington, D.C.: U.S. Committee for Refugees, 1998).

28. A total of 500,000 Azeris were internally displaced within their own country at the end of 1997. U.S. Committee for Refugees, *World Refugee Survey, 1998.*

29. *RFE/RL Newsline,* January 21, 1999.

30. Martha Brill Olcott, "How New the New Russia? Demographic Upheavals in Central Asia," *Orbis* 40, no. 4 (Fall 1996): 547.

31. Legislation passed in 1989 made Kazakh the official language of the republic, although the law would not take effect in the most heavily Russian areas for 15 years. A new language law was passed in 1997 extending the date by which all citizens would be expected to function in Kazakh to January 1, 2006. This law purported to put Russian on an equal footing with Kazakh through such provisions as requiring both languages to be used in the armed forces, police, and security services. Russians worry, however, about provisions of the law which allow Kazakh proficiency tests to be given to applicants for certain administrative, managerial, and service-sector posts and which require at least half of all television and radio broadcasts to be in Kazakh. For more information on the effect of these language laws, see William Fierman, "Formulation of Identity in Kazakhstan's Language Policy Documents 1987–1997" (paper presented at the Association for the Advancement of Slavic Studies Conference, Seattle, Wash., November 1997).

32. During that period, 923,000 Russians left Kazakhstan, and only 213,000 moved into the country. Galina Vitkovskaya, *Emigration of the Non-Titular Population from Kazakhstan, Kyrgyzstan, and Uzbekistan* (unpublished manuscript, 1998), 13–14.

33. Nadezhda M. Lebedeva, *Novaia Russkaia diaspora (The New Russian Diaspora)* (Moscow: Institute of Ethnology and Anthropology of the Russian Academy of Sciences, 1995), 51. Multiple motivations were accepted.

34. *Rossiiskaia gazeta,* April 23, 1997, p. 7.

35. Ludmila Maksakova, "Principal Characteristics of the Migration Situation in Uzbekistan," in *Migration Situation in the CIS Countries,* ed. Zhanna Zayonchkovskaya (Moscow: CIS Research Center on Forced Migration, 1999), 237–238.

36. Lebedeva, *Novaia Russkaia diaspora,* 88–89.

37. Sabit Bagirov, "Sotsial'no-ekonomicheskie aspekti perekhoda k rinku v Azerbaijane" (unpublished paper, Baku, Azerbaijan: January 1999).

38. According to United Nations Development Program figures for 1997, ethnic Kazakhs made up 50.6 percent of the population, with Russians making up 32.2 percent and other nationalities 17.2 percent. United Nations Development Program, *Human Development Report. Kazakhstan 1998* (Almaty: UNDP, 1998), 77.

39. Khalq Sozi, March 17, 1999, in *FBIS Daily Report,* SOV-1999-0322, March 23, 1999.

40. This fact became known when Kazakhstan announced the reopening of 162 primary and secondary schools, 17 preschools, 29 vocational-technical schools, 20 night schools, and 4 boarding schools. No information was released on how many schools remained closed. *U.S.-Kazakhstan Business Council News Wire,* May 5, 1998.

41. Red Cross, *Kazakhstan Vulnerability Survey* (November 1997). According to this source, the total number of children who did not attend school comprised 34 percent of the country's children; 46 percent of the entire population (adults included) did not have winter shoes.

42. By late 1999, Ashkhabad and Baku finally appeared to be working out their disagreement amicably.

43. As of 1996, Turkmenistan withdrew about 22.8 billion cubic meters of fresh water a year, and Uzbekistan withdrew about 82.2 billion cubic meters. World Bank, *World Development Report 1998/99* (Washington, D.C.: World Bank, 1999), 207.

44. Construction of Toktogul Dam and Reservoir was completed in 1975. Four other constant volume hydropower facilities, along with Toktogul, compose the Naryn-Syr Darya Cascade. The other dams were constructed between 1962 (Uch-Kurgan) and the mid-1980s (Kurpsai, Tashkumyr, and Shamaldysai).

45. Vladimir Berezovskiy: "'Water Bomb' Over Central Asia," *Rossiiskaia gazeta,* August 2, 1997, in *FBIS Daily Report*, SOV-97-218, August 6, 1997.

46. Peter Sinnott, "Central Asia's Geographic Moment," *Central Asian Monitor,* no. 4 (1997): 30.

47. Roland Eggleston, "Uzbekistan: Conference to Review Environmental Dangers," *RFE/RL Newsline,* September 18, 1998.

48. By the late 1970s, no water from the Syr Darya was reaching the Aral. Michael H. Glantz, "Creeping Environmental Problems in the Aral Sea Basin," in *Central Eurasian Water Crisis: Caspian, Aral, and Dead Seas,* ed. Iwao Kobori and Michael H. Glantz (Tokyo: United Nations University Press, 1998), 45.

49. Glantz, "Creeping Environmental Problems," 25.

50. Vladimir Akimov, "Central Asian, US Peacekeepers to hold joint exercise," *ITAR-TASS,* March 11, 1997.

51. Anatoly Yurkin, "Large-scale international military exercise in Central Asia," *ITAR-TASS,* September 16, 1997.

52. Yury Chernogayev and Boris Volkhonsky, "NATO Paying for CIS Defense," Kommersant-Daily, September 23, 1998, p. 4, in *Current Digest of the Post-Soviet Press,* October 21, 1998.

53. Of course, if promises of vast energy wealth materialize, these countries could choose to invest in defense, but such promises remain decidedly hypothetical.

54. *ITAR-TASS,* January 4, 1993, in *FBIS Daily Report*, SOV-93-002, January 5, 1993.

55. For more information on these negotiations, see Sergei Gretsky, "Regional Integration in Central Asia," *Analysis of Current Events* 10, no. 9–10 (September–October 1998): 12–13.

56. On February 11, 1999, for example, the Kazakh Ministry of Energy, Industry, and Trade announced that, effective March 11, it would introduce tariffs of up to 200 percent on foodstuffs and certain other imported products from both Kyrgyzstan and Uzbekistan. The ministry cited the need to protect Kazakhstan's producers from the inexpensive, subsidized imports that come out of those countries. Bruce Pannier, "Central Asia: Concern Grows Over Possibility Of Trade War," *RFE/RL Newsline,* February 16, 1999.

57. Bruce Pannier, "Central Asian States Pledge 'Eternal Friendship'," *OMRI Analytical Briefs* 1, no. 523 (January 14, 1997).

8

U.S.-Iranian Relations: Competition or Cooperation in the Caspian Sea Basin

Geoffrey Kemp

For the past few years there has been growing interest in the Caspian Basin, primarily because of the belief that the region contains huge deposits of petroleum and natural gas. A related factor is that with the breakup of the Soviet Union, three of the newly independent countries who own much of the anticipated oil and gas reserves—Azerbaijan, Turkmenistan, and Kazakhstan—are welcoming foreign energy companies and offering them lucrative investment opportunities.

As a consequence, there has been talk of a "New Great Game" and a "Black Gold Rush." Over $70 billion has been earmarked for investment in developing the energy, and a world class battle is being waged by neighboring countries to provide egress routes for the oil and gas to reach world markets. Russia, China, the United States, Turkey, Georgia, Iran, Pakistan, and Afghanistan are all putting forward proposals for pipeline routes that best suit the interest of the region and the world energy market. As the decisions over preferred routes reach a climax, other geopolitical issues have become embroiled in the issue, including the civil war in Afghanistan, the conflict between Armenia and Azerbaijan over Nagorno-Karabakh, succession movements and fighting in Georgia and Russia, Turkey's enduring conflict with the Kurds, and Iran's continuing confrontation with the United States. In addition to ongoing conflict, the region is beset with political instability, corruption, and economic mismanagement.

This menu of potential wealth alongside present problems raises a very basic question: How stable is the Caspian region and how realistic are the prospects that it will become a major player in the world energy market? The reality is that despite much hype and enthusiasm about the Caspian's promise, the politics of the region may ruin whatever hopes there are for an open-ended bonanza.

In assessing the Caspian's importance for the Persian Gulf energy producers, it is necessary to review the very complex geopolitics of the region
and the wildly different interpretations of the facts concerning energy
development in Central Asia and the Caucasus. Consider the following
statement, which is typical of some of the euphoric writing coming from
journalists who visited Baku and other oil regions of the Caspian in the
late 1990s:

> The region's wells will shower unimaginable wealth on people whose annual per
> capita gross domestic product (GDP) today hovers between $450–$600, build-
> ing a new El Dorado in nations where camels still outnumber automobiles in
> 1998. . . . With many of the world's current principal oil sources drying up, the
> Caspian reserves are vital to the continuing prosperity of the West, and the
> advancement of the Western world.[1]

Contrast this statement with the words of a well-respected Central Asia
specialist at the Carnegie Endowment for International Peace, Martha Brill
Olcott:

> The most grandiose oil and gas projects are still in their early stages, with their via-
> bility and pace of development uncertain. The reality of post-communist devel-
> opment has instead been an increase in corruption and a sharp drop in living stan-
> dards once protected by a comprehensive welfare net. Growing poverty has made
> a burgeoning trade in narcotics and illegal arms—other "global" commodities for
> which this part of the world is known—that much harder to stamp out; it has also
> rendered the population more susceptible to the appeal of Islamic radicalism. And
> ironically, many of the oil-rich region's leaders have taken the recent strategic
> blandishments of powers such as the United States as proof that their own posi-
> tion will be protected from either internal unrest or outside attack.[2]

Since there is now a growing literature on the politics of Caspian energy,
this chapter will not attempt to review the entire spectrum of issues. Rather
the focus will be on the role of Iran as both a Caspian and Persian Gulf country and its relations with both the United States and Russia.[3] It will be argued
that the nature of this relationship will have a strong influence on the debate
about Caspian energy and its impact on the Persian Gulf. First, however, it is
necessary to make some general points about Caspian energy and the status
of offshore ownership.

HOW IMPORTANT IS THE CASPIAN?

The Caspian Basin is one of several *potentially* massive energy regions that
could be developed over the next twenty years. For years, energy specialists

have known that eastern Siberia contains enormous oil potential, but little so far has reached the international market. A combination of politics, economics, and the environment have slowed development. Likewise in China's western desert, the Tarim Basin once was believed to contain huge deposits, but so far the problems of egress and financing have been overwhelming and the early exploratory efforts have been disappointing. The South Atlantic, off Argentina, may be rich in oil but the dispute over ownership of the Falkland Islands is likely to delay efforts to explore and develop the region's oil potential.

In the case of the Caspian, what is known is that at least 30 billion barrels of oil are proven to be available—far short of the 200-billion-barrel figure often cited. The latter figure could be correct once the required drilling has taken place, but drilling requires huge rigs that have to be transported over excruciatingly difficult routes. For much of 1998, there was only one large exploratory semi-submersible drilling rig in the Caspian.[4] Furthermore, while at this time $70 billion of investment has been pledged by foreign oil companies, not much more than $4 billion has been spent.

Three dimensions of the Caspian energy equation are of concern: (1) the ownership of the resources; (2) the amount of hydrocarbons available for extraction; and (3) the production and distribution costs, including direct and environmental costs. Whether there will be cooperation or conflict over access to Caspian energy will depend upon whether these issues are resolved in an adversarial or cooperative manner.

Until 1991, the Caspian Sea was controlled by the Soviet Union and Iran. The final legal status of the Sea following the break-up of the Soviet Union has yet to be defined. Moreover, all five of the littoral states face major economic and political obstacles limiting their capacity to export energy. Three of the producers—Azerbaijan, Kazakhstan, and Turkmenistan—are surrounded by other countries. They cannot get their energy to market without crossing someone else's territory. Interestingly, in the long and turbulent history of the oil business this has rarely happened. In the past, all the great oil and gas exporting countries have had direct access to the world's shipping lanes. (Over the years the major oil and gas exporting countries have included: Algeria, Angola, Bahrain, Brunei, Canada, Columbia, Egypt, Iran, Indonesia, Iraq, Kuwait, Libya, Mexico, Nigeria, Norway, Qatar, Oman, Romania, Russia, Saudi Arabia, United Arab Emirates, the United Kingdom, the United States, and Venezuela.)

Owing to this geographic reality, the multiple routes being considered to transport the energy have assumed political and strategic, as well as economic and environmental, overtones. However, since the economic stakes for the winners are so high, the key participants are only reluctantly accepting the need to cooperate and compromise. Russia, for instance, has tried to use strong-arm tactics to assure a monopoly of the key routes from Azerbaijan and Kazakhstan. However, its ability to dictate terms has been limited because of its own financial problems and the chaos and conflict along its southern borders. Iran's problems relate to its poor relations with the United States, as well as its difficulties attracting the investment needed to develop its huge but underutilized natural gas reserves and to exploit its geography as a transit route for its neighbors' oil and gas.

Since the choices for oil and gas routes involve so many complicated decisions, many different options have been proposed involving countries as far apart as Bulgaria, Greece, and Turkey in the west, and India, China, and Pakistan in the east. Each of the many options has costs and benefits. For instance, Bulgaria and Greece, in cooperation with Russia, have proposed a route for oil transportation that would avoid Georgia, Turkey, and the Bosporus Strait, but the plan would probably further aggravate Greek-Turkish relations. One of the more intriguing proposals has been to build natural gas pipelines from Turkmenistan and Iran that run into Pakistan and then on into India.

THE LEGAL STATUS OF THE CASPIAN SEA

The legal debate over the Caspian Sea traces back to the 1921 Treaty of Moscow, reaffirmed in 1935, which declares that the inland Caspian Sea belongs to Russia and Persia.[5] Later, the Protocol of 1940 superseded the Treaty and made the Caspian joint Soviet-Iranian property.[6] In 1960, Iran and the Soviet Union agreed to arbitrary lines establishing territorial waters in the Caspian. Russia's basic position since 1991 has been that the legal status of the Caspian is based on the earlier treaties between the Soviet Union and Iran and is not covered by the 1982 UN Convention on the Law of the Sea (UNCLOS) since the Caspian is in fact a "special inner sea" and has no natural connection with any other "sea."

In contrast, Azerbaijan and Kazakhstan have argued that the Caspian Sea should be treated under Article 122 of UNCLOS, which defines a semi-enclosed or enclosed sea as "a gulf, basin or sea surrounded by two or more states and connected to another sea or ocean by a narrow outlet or consisting entirely or primarily of the territorial seas and exclusive economic zones of two or more coastal states."[7] Under this definition the Caspian would be covered by the latter qualification. Furthermore, there are precedents for dividing large lakes between two or more states (e.g., Lake Victoria, Great Lakes of North America, and Lake Titicaca).[8] Russia has argued that the Caspian legally should be treated as joint property shared by all the bordering states— including the newly-independent republics, Russia, and Iran—which would effectively delegitimize efforts by any one state to unilaterally exploit the Sea's resources within its own demarcation lines. The demarcation would be drawn to equidistant lines between bordering states to effectively divide up the entire sea.

Recently, however, Russia has adapted a more conciliatory tone and has sided with Turkmenistan, Kazakhstan, and Azerbaijan regarding the demarcation of the Caspian Sea. This has left Iran more isolated. While Moscow already has signed a non-binding protocol with Azerbaijan and was expected to conclude a treaty with Kazakhstan, Iran's argument remains that it will accept unilateral exploitation of the Caspian Sea's oil and gas reserves only if the sea is sectioned into equal sectors.

Iran has a strong interest in claiming a share of the seabed beyond its own sector as defined by UNCLOS because the main oil fields lie in the middle of the sea, off Azerbaijan. The least promising waters are those off Iran. Until there is full agreement between the littoral states on the demarcation issue, it can continue to be a source of friction and become another factor to frustrate efforts to explore fully the potential of the offshore region.

Disputes over the Caspian are paralleled by an age-old dispute between Turkey and Russia over the Turkish straits, which continue to comprise a critical export route for energy resources from the former Soviet republics. The 1936 Montreux Treaty bars Turkey from restricting Russian merchant shipping through the Bosporus Strait. However, it appears that Turkey is prepared to mobilize environmentalists and world opinion to limit Russian oil transit on the waterway and will promote its own alternative for bringing Caspian oil to the Mediterranean Sea via the proposed Baku-Ceyhan pipeline, which would bypass the Black Sea.[9] There is uncertainty about how far Turkey will go to limit tanker traffic. Experts point out that in 1996 two million tons of oil flowed eastward through the Bosporus Strait from the Mediterranean Sea to Bulgaria and Romania. This figure could drop considerably when Caspian oil comes on line and can be shipped to these countries without transiting the Strait.[10] There is no doubt that Turkey's environmental concerns have merit, as anyone who has crossed the Bosporus can attest. However, the environmental issue is only one factor in the complicated competition for access routes.

THE IMPORTANCE OF IRAN

While Russia is clearly the most powerful of the Caspian littoral states, Iran is a key player with a significant impact on Caspian geopolitics and economics. How the United States and Iran resolve (or do not resolve) their problems is a significant factor influencing the future direction of both Caspian Basin and Persian Gulf stability. A rapprochement between Washington and Tehran would permit economic, rather than political, criteria to determine pipeline routes and investment opportunities. A marked deterioration in U.S.–Iranian relations could have a most negative impact on regional development in the Caspian region and would likely increase the prospects for violence and instability. Iran's relations with Russia and its Caspian neighbors will be addressed first, followed by a discussion of U.S.-Iranian relations.

Iran and Russia

Iran and Russia are not only the two most powerful Caspian countries, they also share many common interests in other parts of the greater Middle East. They both regard long-term instability in Central Asia and the Caucasus as inimical to their security and have worked carefully to ensure that the other regional countries remain either stable or within their respective spheres of influence. When instability has occurred, as in Chechnya and Georgia, they have gone to great lengths not to antagonize each other. Both countries stand to benefit from projects to develop the potential energy wealth of the Caspian Basin. As noted, they still have outstanding differences on how to demarcate the offshore resources of the Caspian, with Iran's position more at odds with the smaller countries than with Russia.[11] However, both disapprove of the efforts being made by the United States to determine the selection of pipeline routes, especially the proposed trans-Caspian routes that would bypass both Russia and Iran. They share a common perception that the United States is trying to minimize their influence in a region of which they are a part.

Absent a major change in U.S.-Iran relations, it is likely that this alliance of interests will continue. Iranian policy toward the countries of the former Soviet Union has been pragmatic and devoid of the ideology associated with its activities in other regions, including the Persian Gulf, Lebanon, and Africa. Nevertheless, over time there are potential conflicts of interest between these two powers. Since both Iran and Russia are natural egress routes for Caspian oil and gas, if Iran and the United States were to repair their relationship and Iran were permitted to become a key egress route, it would run into direct competition with Russia. This competition will be especially intense if production levels of the Caspian oil and gas fields fall short of some of the more optimistic estimates. Even with its current system of pipelines, Iran claims that it could shift up to 700,000 barrels per day in the form of swaps from the Caspian countries through its domestic pipeline network, exporting its own oil from the Persian Gulf while using Caspian oil for its own refineries and its market in the north. Although this figure is probably exaggerated, there is no doubt that oil swaps are likely to become a significant factor in Caspian exports and that Iran will benefit from this arrangement.

Similarly, a major buildup in Iranian military capabilities, particularly if it involves long-range missiles and weapons of mass destruction, could eventually pose a threat to Russia. The testing of the Shehab-3, developed with Russian assistance, demonstrates that Iran could eventually possess the capability to reach parts of Russia. Hence the question arises, why is Russia helping Iran in its military acquisitions? The answer is to be found in the confused and conflicting state of affairs in Moscow. As with other foreign policy issues, including relations with China, Russian policies seem to contradict each other. The Ministries of Foreign Affairs and Defense are often at odds with aggressive lobbies pushing for arms sales and technology transfers. Moreover, Russia's huge oil and gas companies, especially LUKoil and Gazprom, have great clout in Moscow and in many ways operate their own foreign policies. During his tenure as prime minister, Evgenii Primakov, attempted to impose some discipline and a more coordinated strategy on the activities of the various Russian interest groups that deal with Iran, but it is not clear how successful this effort has been. A number of individual Russians and small Russian companies and institutes have been aiding Iran with its missile and nuclear energy programs. When approached by American officials, the Russians deny that there is any formal government policy in favor of such help. Missile cooperation, in particular, would be a violation of the Missile Technology Control Regime (MTCR), which Russia has signed. Yet Russia seems to be incapable or unwilling to effectively enforce its own laws—which either reflects the weakness of Moscow or a malevolent attempt to undermine American interests by building up Iranian capabilities. After much U.S. pressure and threat of sanctions, it was reported on March 17, 1999, that Russia had agreed to curtail its support for some of Iran's nuclear activities, but not including its plan to build nuclear power reactors for Iran at Bushire.[12]

Iran and Turkey

Iran and Turkey have an important, but ambivalent, relationship. There have been allegations of Iranian sponsored terrorist attacks against Turkey's secular intellectuals, journalists, and politicians. From June 1996 to June 1997, when Necmettin Erbakan held the Turkish premiership, relations between the two countries improved in view of Erbakan's Islamist credentials and his bitter criticisms of the West and Israel. However, since his ouster, political relations have become more tense, especially in light of the security cooperation between Turkey and Israel. The relationship took a turn for the worse in early 1998 when both countries recalled their ambassadors for several months. Moreover, Turkey continues to be concerned about Kurdish rebels in the southeast and has repeatedly accused Iran of supporting its outlawed Kurdish Workers Party (PKK). During the Turko-Syrian crisis of autumn 1998, also over alleged support for the PKK, Tehran repeatedly pledged its support for Damascus during Turkey's military buildup opposite the sparsely defended stretches of northern Syria.

Nevertheless, the key to Iranian-Turkish relations may be economic. With a growing population of 62 million, Turkey expects serious energy shortfalls in the near future. Estimates show that energy demand will rise by 200–300 percent in the next 10–15 years, and Iran may become an important supplier.[13] A proposed gas pipeline through Iran to Turkey that would bring first Turkmen and then possibly Iranian gas to Turkey's huge market may still become a reality in view of the economics and Turkey's reluctance to become overly dependent on Russia for energy supplies.

Turkey's cooperation with Iran on energy projects runs counter to preferred U.S. policy, which is to support an "east-west" option to bring both oil and gas from Azerbaijan, Kazakhstan, and Turkmenistan through Georgia to Turkey. This route would achieve the Clinton administration's goal of limiting major egress routes through Russia and Iran, as discussed in more detail below.

Iran's Role in the South Caucasus and Central Asia

Iran's proposals to cooperate with Turkey, Azerbaijan, Kazakhstan, and Turkmenistan on energy-related projects have complicated U.S. efforts to isolate Iran and to deny it participation in Caspian projects, including pipeline routes. The net effect of U.S. pressure has been to make it very difficult for the littoral countries to establish serious cooperative ventures with Iran.

In this regard, Iran's attitude toward its neighbors in the South Caucasus is complicated. Georgian President Eduard Shevardnadze's regime remains threatened with ongoing violent clashes in the north. Iran's relations with the former Soviet republic have traditionally been fairly cool, but some steps towards reconciliation between the two countries were made in 1998. Stressing that cooperation would strengthen security and tranquillity in the region, Iranian President Ali Mohammad Khatami called for an expansion of relations in a meeting with Georgian Foreign Minister Irakli Menagarishvili.[14] Menagarishvili's visit to Tehran resulted in a memorandum of understanding that stressed the importance of setting up a joint economic commission between the two countries. The two sides also emphasized the need for peaceful resolution of regional disputes, including in Georgia's breakaway Republic of Abkhazia, and the importance of safeguarding security in the Caucasus.[15]

Tehran has not approached Georgia's western neighbor, Azerbaijan, in the same way. Despite its potential oil riches, Azerbaijan faces deteriorating economic and social conditions. The conflict with Armenia over Nagorno-Karabakh, whose population is overwhelmingly ethnically Armenian, remains unresolved, and Iran continues to provide economic support to Armenia.

From Tehran's point of view, Baku's close relations with the United States and its stance on the demarcation of the Caspian Sea remain obstacles to warmer relations between the two neighbors. On the issue of pipelines, Azerbaijan's president, Heydar Aliyev, is in favor of the Baku-Ceyhan route advocated by the Clinton administration. The Azerbaijani government has accused Iran of financially supporting a coalition of opposition parties in Azerbaijan as well as the banned Islamic Party of Azerbaijan (IPA), and four IPA leaders have been imprisoned on charges of espionage for Iran.

In addition, Iran is also struggling with its eastern neighbor, Afghanistan. Tehran opposes the Taliban regime, which has attempted, hitherto with little success, to establish complete control over the entire country while it continues to impose its Sunni ultra-fundamentalist interpretation of Islam over the part of the country it governs. The northern enclave, small but well-armed (mainly by Russia), eluded the Taliban's grip for some time, but the northern opposition lost a major stronghold with the fall of Mazar e-Sharif in the summer of 1998. The regime has no bureaucracy, no educated cadre, and not a single trained economist, engineer, or oil expert.[16] At the same time, Afghanistan faces physical, economic, and social devastation from years of civil war.[17]

Following the fall of Mazar e-Sharif, it came to light that the Taliban had killed a great many Shiite Afghanis, people who traditionally looked to Iran for support and protection, and had kidnapped more than forty Iranian nationals. The Taliban failed to respond to demands to turn over the Iranians, some of whom had been murdered, and Iran proceeded with a buildup of some 200,000 troops on its border with Afghanistan under the guise of military exercises. War was averted when the Taliban released the Iranians, but only after bombastic threats from Iran and responses from the Taliban that included threats to use its Scud SSMs against Iran.

On the other hand, Iran has established a good relationship with Afghanistan's common neighbor, Turkmenistan. This gas-rich Caspian state, under the authoritarian leadership of President Saparmurat Niyazov, is also facing an uncertain future. Economic deprivation of the vast majority of the population, worsened by burgeoning corruption, could cause social unrest if the oil dividends fail to benefit the population as a whole. Niyazov's unwillingness to enact any market reforms has translated into continued negative growth and high inflation. Under these conditions, obtaining income from exports of its natural resources is absolutely essential.[18]

Iran has helped relieve some of Turkmenistan's economic troubles. The Korpeje-Kurt-Kui gas pipeline from Turkmenistan to Iran, put into operation in December 1997, has now become the only export pipeline from Turkmenistan not dependent upon Russia.[19] In addition, a visit in July 1998 by Niyazov to Tehran resulted in the signing of a joint communiqué and three agreements on cooperation in the fields of politics, economics, and sports.[20] In the communiqué, Iran and Turkmenistan voiced objection to the agreement between Russia and Kazakhstan dividing the northern sector of the Caspian Sea.[21]

However, future energy cooperation between the two countries received a severe jolt in February 1999 when it was announced that Turkmenistan had signed a deal with two American companies, General Electric and the Bechtel Group, to lay a gas pipeline across the Caspian seabed to Azerbaijan and into Turkey through Georgia. If built, this pipeline would obviate the need for a pipeline to Turkey via Iran.[22]

The demarcation of the Caspian also continues to complicate Iran's relationship with Kazakhstan, the largest Central Asian country. Like many of its neighbors, Kazakhstan, led by President Nursultan Nazarbaev, is battling deteriorating economic and social conditions. The president has heightened popular resentment by spending billions of dollars on a new capital in Akmola, now renamed Astana.[23] The country's international relations are made intricate by its need to perform a balancing act between China and Russia while at the same time being vulnerable to Russian pressure because of its 37 percent Russian minority.

Despite U.S. pressure, Kazakhstan's leadership has indicated that it wants to develop relations with Iran. In November 1997, the Kazakh ambassador to the United States, Bolat Nurgaliyev, emphasized the importance for Kazakhstan of trade and economic contacts with Iran.[24] During a visit to Tehran in March 1998, Kazakh Foreign Minster Kasymzhomart Tokayev voiced the Central Asian republic's readiness for cooperation in the oil sector.[25] In November 1998, the Kazakh Energy Ministry reiterated its approval of oil swaps with Iran.

IRAN'S ENERGY DILEMMA

Iran's relations with its Caspian neighbors are strongly influenced by its complicated energy situation, its unique geography, and its continuing conflict with the United States. In order to develop its economy to meet the growing expectations of a large young and dissatisfied population, Iran must undertake a massive capital investment in its energy sector and work closely with both its energy-rich and energy-deficient neighbors to achieve a mutually beneficial economic relationship. Mismanagement of the Iranian economy by the regime has compounded the problems to the point where even if U.S.-Iran relations were to improve, U.S. sanctions were removed, and objections to pipeline routes were laid aside, it is questionable whether foreign companies would invest significant sums of money in Iran until its government enacts far-reaching reforms designed to protect the interests of foreign capital. This would entail opening up its oil and gas sectors to foreign investment, a step that would be regarded as anathema by the more die-hard conservatives—an integral component of the revolution in its early days—who remain opposed to foreigners and their money. Although Iran has already taken steps to involve foreign companies—most obviously in the case of the aborted Conoco deal and the follow-up agreement with Total, Gazprom, and

Petronas—so far the concessions have been restricted to offshore energy assets.[26] Undoubtedly, a way will be found to get around Iranian law (which forbids foreign investment in onshore fields), primarily by negotiating "buy backs."[27] Nevertheless, until the investment climate has been tested and foreign companies are assured that they will be treated fairly, the likelihood of massive investment seems slim, despite increasingly friendly overtures from the Iranian government.[28]

A few facts on Iran's energy situation help to explain why this issue commands such a priority in Iranian politics. Iran remains one of the world's largest oil producers. Revenues fell significantly in recent years along with production, domestic consumption has increased, and the base price of oil on the world market dropped to less than $9 a barrel in 1998. However, since then 1999 oil prices have risen, providing Iran with much needed new revenue.

In 1978, the last full year of the Shah's rule, Iran produced 4,252,000 barrels of oil per day (b/d). Of this, domestic consumption accounted for 517,000 b/d or 12 percent of total production, leaving 3,735,000 b/d for export. In 1997, production amounted to 3,730,000 b/d, of which 1,255,000 b/d or 33 percent was for domestic consumption, leaving 2,475,000 b/d for export. Thus, not only has production fallen but the revenues from exports as a percentage of total production have fallen from 88 percent to 67 percent.[29]

Because of serious production and maintenance problems, Iran is working to rebuild its deteriorating oil fields.[30] Production in 1997, although lower than in 1978, actually increased from levels in the early 1990s. From 1993 through the end of 1996, Iran boosted production to an average of 3.6–3.7 million barrels of oil per day, pumping as much as 4 million barrels on some days to demonstrate productive capacity.[31] Some have suggested that such high levels of pumping will damage Iranian oil reserves. Indeed, recent reports suggest that Iran has been unable to meet its OPEC quota. At these high levels of production, Iran's proven oil reserves will last just over forty years. The depletion or damaging of Iran's existing reserves makes expansion of proven reserves essential. As part of its effort to get a larger share of the resource pie, Iran has attempted to be part of deals in the Caspian Sea.

Iran's other great asset is natural gas, regarded globally as one of the most desirable sources of energy because of its abundance, competitive production costs, and environmental cleanliness. Yet from the perspective of the Persian Gulf countries, which all have huge proven reserves, significant revenues from natural gas exports are very much a part of their future rather than their present economic calculations. Although Iran contains the world's second largest gas reserves (after Russia) with nearly 16 percent of the world's total, its exports of gas are minuscule, with only 0.1 billion cubic meters sold to Azerbaijan in 1997. The remainder is used for domestic consumption and reinjection into aging oil wells

Iran's energy dilemma to raise pressures necessary to bring the oil to the surface is compounded by its demographic distribution, which determines domestic demand, and the actual location of its oil and gas deposits and foreign markets. The majority of Iran's oil and gas wealth is located in the south and southwest of the country, whereas the majority of the Iranian population (over 65 million total) is located in the big cities in the north. Most of Iran's oil markets are in Europe and Asia, and, aside from Turkey, its foreign gas market potential is in South Asia.

Because of the asymmetries between the location of the energy and the markets, it makes economic sense to bring oil and gas from Iran's northern neighbors to meet Iran's northern demand and to use Iran's oil in the south for distribution elsewhere, most notably to the Asian markets and, in the case of gas, to India and Pakistan.

One of the most promising developments from the Iranian point of view is the notion of swapping oil from Kazakhstan, Turkmenistan, and possibly Azerbaijan and Uzbekistan for use in the Iranian market to be offset by equivalent amounts of Iranian oil from its southern fields to be shipped to markets in Europe and Asia. While there are complicated financial and technical matters to work out, including tariff costs and the compatibility of crude oil from the Caspian fields with Iranian refineries, Iran's demand for petroleum is sufficiently great that it can absorb large quantities of oil from its northern neighbors provided that it can be delivered and processed at reasonable costs. Thus, Iran has proposed building a new pipeline from its Caspian port of Neka to Tehran to link up with the Iranian pipeline network, which would permit Caspian oil to proceed to refineries at Tehran, Tabriz, Arak, and various other locations in the north. This petroleum would then be refined into gasoline, kerosene, and other products that the Iranian market could easily absorb. Oil that presently comes from the south to the north to meet Iran's domestic needs could then be exported to foreign markets, and the receipts from these additional exports would be repaid to the northern producers. Another proposal would be to bring oil from Baku to Tabriz for use in the Iranian market or, if politics permit, to be shipped through a longer pipeline extending into Turkey.

Like all issues relating to Caspian trade, much depends on the logistical feasibility of such movements. Oil can, after all, be moved by many methods from the producing states to the Caspian port of Neka—barge, ship, pipeline, and train. Given the dire straits of the economies in the former Soviet republics, any short-term additional revenue will be extremely important, which makes the Iran option advantageous. It will take years to complete the trans-Caspian pipeline, the northwestern route from Tengiz to Novorossiisk, or the eastern route to Asia. If, in the meantime, the Iranians can begin to swap oil with their neighbors, it could promote investment in Iran's oil-transportation infrastructure that would eventually raise the number of barrels swapped to between 400–500 thousand per day. (Some suggest the figure could go as high as 700,000 b/d.) This is not a negligible amount of oil and could compete with oil that would otherwise traverse different routes, including the trans-Caspian, Russian, or Chinese routes. Furthermore, the Iranians argue, the costs of this proposal are lower than the costs of the proposed east-west pipeline. However, it would add to the traffic flowing through the already crowded Persian Gulf and Strait of Hormuz, raising for some shippers the strategic dangers of dependency on this egress route.

U.S. POLICY ON IRAN AND PIPELINES

Since the Iranian Revolution in 1979, successive American administrations have pursued a tough and often confrontational policy towards the Islamic Republic. Indeed, it can be argued that though the collapse of the Soviet Union was a bonus for U.S. foreign policy in most areas of the world, in the case of Iran it has made relations more difficult. U.S. support for the new republics of the Caucasus and Central Asia has drawn Russia and Iran closer together to challenge what they regard as American hegemonic aspirations. Several factors inhibit improved U.S.-Iran ties. Iran's anti-Israel policy is a major obstacle to normalization. As long as Tehran pursues policies that directly threaten Israel, no U.S. administration nor the U.S. Congress will initiate or accept radical changes in American policy toward Iran. Although President Khatami's election in May 1997 has radically changed the dynamics of Iranian domestic politics, unless he and his moderate supporters gain control of the key instruments of power, his proposed reforms and even his tenure in office could be in jeopardy. Unfavorable economics and a young and increasingly dissatisfied population in Iran pose serious challenges for any Iranian leader, whether moderate, centrist, or radical. The negative impact of American sanctions on Iran's vital energy sector provides a strong incentive for the regime to improve relations with the United States. However, because America remains a dominant factor in Iran's strategic, political, economic, and psychological ethos, the regime's conservatives realize that rapprochement will inevitably mean the diminution, if not the end, of their power.

On June 17, 1998, Secretary of State Madeleine Albright called for the United States and Iran to build greater mutual trust so Washington could "develop with the Islamic Republic, when it is ready, a road map leading to normal relations."[32] The official Iranian response has been muted and cautious. Critics of the Clinton administration argue that helping Khatami to open up Iranian society will undermine the moderates by stimulating a conservative backlash. Whichever view prevails, there will be no real progress until six key issues are addressed: historic mutual grievances; terrorism; the

U.S. military presence in the Persian Gulf; Iran's weapons programs; its opposition to Israel and the peace process; and economic issues, especially sanctions and pipeline policies.

This last factor is especially contentious. Despite its proclaimed policy of promoting multiple pipeline routes to bring oil and gas from the Caspian Basin, the Clinton administration is committed to "avoiding transit routes through Iran."

The United States has a major interest in Caspian oil and gas for a number of geopolitical and economic reasons, including its growing dependency on oil imports over the next decade. Thus, like all industrial powers, the United States wants to ensure against energy supply disruptions and seeks international cooperation on global energy issues.[33] It follows from these overall objectives that the core of the Clinton administration's energy policy in the Caspian region is the multiple pipelines strategy. According to former Energy Secretary Federico Peña, multiple pipeline routes "will enhance the commercial interests of companies operating in the Caspian states as well as host states." However, the United States does not want Russia unilaterally to decide the rules of the Caspian oil and gas opportunities, and it wants to exclude Iran from the game. It will be impossible to exclude Russia since the Caspian Pipeline Consortium's pipeline from Tengiz to Novorossiisk has already begun construction. Thus, the United States is pressing for an "east-west energy corridor," i.e., a Baku-Ceyhan line with a trans-Caspian extension to Turkmenistan and Kazakhstan. Such a pipeline would supply extra volumes of oil to justify the high cost of Baku-Ceyhan. The Clinton administration is also proposing a trans-Caspian gas pipeline to bring Turkmen gas to Turkey via Azerbaijan, thereby bypassing Iran.

In an April 1998 statement to the House Committee on International Relations, Secretary Peña succinctly reviewed U.S. policy in the Caspian region:

> The Administration is committed to an "east-west energy corridor" to link the countries in the region to the west. Our east-west energy corridor initiative principally supports the construction of trans-Caspian oil and gas pipelines running under the Caspian Sea, linking the countries on the eastern Caspian shore—Kazakhstan and Turkmenistan—with those on the west, starting with Azerbaijan in the coastal city of Baku. Both lines would traverse Georgia. The oil line would proceed to the Mediterranean port of Ceyhan, while the gas line would serve Turkey's growing gas market.[34]

Although the Turkmen government in February 1999 signed an agreement with a U.S.-led consortium to build a 2,000-kilometer pipeline under the Caspian to Turkey, there remain overwhelming political and environmental problems, and the economic viability of the trans-Caspian route is doubted by many. In order for the project to be worthwhile, it would only make sense if it included both oil and gas and if it took the majority of oil and gas produced by Kazakhstan, Uzbekistan, and Turkmenistan across the sea into

Azerbaijan. This would most likely work out if the northern route to Russia and Novorossiisk or the southern routes through Iran or Afghanistan do not materialize. Thus, the irony is that in developing a plan for a trans-Caspian pipeline, the United States is undermining the viability of other multiple egress routes, which is counter to its original strategy. Furthermore, if the pipelines were built, it would endow Azerbaijan with considerable strategic importance since it would be the junction for not only its own oil but also oil and gas coming from its neighbors across the Caspian.

There are other factors that suggest that the United States should reconsider its policy, both in terms of the overall energy equation in the Caspian Basin and in terms of its relationship with Iran. First, reports have questioned the potential of the region. A Baker Institute study said that even under the most optimistic assessments, by 2010 Caspian oil production likely will reach little more than 3.5 million b/d and cover only 3 to 4 percent of expected worldwide use.[35] This is due to a range of geological, logistical, and political challenges, some of which were discussed above. The Baker Institute study concludes that there should be only one main export line. In other words, the multiple pipeline strategy would be unsustainable, unless governments or multilateral institutions were to subsidize them. The International Institute for Strategic Studies in London followed up by stating that the U.S. Department of Energy's projection of around 200 billion barrels of oil in the Caspian Sea is "widely derided" in the oil industry. It said most forecasts show that the region contains between 25 billion to 35 billion barrels.[36] In addition, the low oil prices in recent years made oil companies even less eager to invest in multi-billion dollar pipeline projects. In short, multiple pipeline routes may *not* enhance the "commercial interests of companies operating in the Caspian region," as former Energy Secretary Peña argued. The rise in oil prices in 1999–2000 has had little impact on pending decisions over pipelines. Concern about volumes of reserves outweigh the importance of current market prices.

Second, given rapidly deteriorating social and economic conditions in Central Asia and the Caucasus, the littoral states are eager to obtain their much needed oil income.[37] As noted above, for Kazakhstan and Turkmenistan, the quickest and cheapest way is to go through Iran. Turkmenistan is particularly eager to export its gas and oil to the Persian state.

Third, the arguments for an Iran option are not only logistical, financial, and ecological, but in some ways also geopolitical. In 1998, Iran defaulted on debt payments to international creditors for the first time in four years and was suffering the results of low oil prices and economic mismanagement. Nevertheless, the Iranian regime is more stable than most of the newly independent states of the Caspian region, enjoys a degree of domestic legitimacy, and (the American embargo not withstanding) has seen a general improvement in its international position. A European official summed the situation up by saying, "The U.S. position of excluding Iran is patently untenable."[38]

Yet so long as oil prices remain low and U.S.-Iran relations remain frozen, in part because Russia is seen to be helping Iran with its weapons programs, it is unlikely that there will be any quick fix to Caspian energy conundrums. Nevertheless, unpredictable developments could change the nature of U.S.-Iran relations. Khatami's replacement by a conservative *mullah* would freeze the prospects for détente. New congressional sanctions against the Iranian energy sector would further chill diplomacy. Some developments, such as Saddam Hussein's death, would be viewed positively in the United States, but more ambivalently in Iran. On the other hand, an unexpected breakthrough in the Arab-Israeli peace process could make it much easier for Iran to temper its anti-Israeli rhetoric and, at the same time, reduce its support for radical groups such as *Hizbollah, Hamas,* and Islamic *Jihad.* In the event of such positive breakthroughs, Washington and Tehran could even consider a "grand bargain" that would seek to limit Iran's nuclear and missile programs in exchange for a more cooperative regional security regime. This, in turn, could open the door for U.S.-Iran cooperation on energy projects, including oil and gas pipelines from the Caspian through Iran.

Greater U.S.-Iranian cooperation on energy projects would be a major boost for long-term investment in Iran's oil sector. It would also improve Iran's ability to provide large amounts of oil for the export market, while removing one of the key sources of instability in the region. It would also help to resolve the conflict over Caspian energy routes—provided that Russia feels it would benefit from such a development. This is a significant issue, for while a rapprochement between the United States and Iran would help remove some of the most divisive issues that plague U.S.-Russian relations, it might also increase Russian concerns about its long-term posture in the Caspian. Thus, how the triangular relationship between the United States, Russia, and Iran evolves will likely be the most important strategic factor in emerging Caspian geopolitics.

NOTES

1. Frank Viviano, "Caspian Basin Oil Boom Transforms Entire Area," *Washington Times,* August 15, 1998, A6.
2. Martha Brill Olcott, "The Caspian's False Promise," *Foreign Policy,* Summer 1998: 96.
3. This paper draws upon two recent publications by the author, Geoffrey Kemp. See his *America and Iran: Road Maps and Realism* (Washington D.C.: The Nixon Center, 1998), and "The Persian Gulf Remains the Strategic Prize," *Survival,* Winter 1998–1999: 132–149.
4. Marshall DeLuca, "Caspian Mobile Rig Clubs Pick Up Slack in Availability," *Offshore,* July 1998: 68, 70.

5. John C. Colombos, *International Law of the Sea* (Great Britain: Longman's, 1961), 164.

6. "Iranian Ambitions in Caspian Region Eyed," *Aziia i Afrika Segodnia*, no. 12 (December 1994): 15–18, in *Foreign Broadcast Information Services (FBIS) Daily Report*, March 10, 1995: 1.

7. See Article 122 of the United Nations Convention on the Law of the Sea <http://www.un.org/Depts/los/unclos/part9.html>(February 6, 2000).

8. Clive Schofield and Martin Pratt, "Claims to the Caspian Sea," *Jane's Intelligence Review*, February 1996: 77–78.

9. See Sabri Sayari, "Turkey, Caspian Energy, and Regional Security," in this volume, for additional analysis of Turkey's position on pipeline routes and tanker traffic through the Bosporus Strait.

10. Schofield and Pratt, "Claims to the Caspian Sea," *Jane's Intelligence Review*, February 1996: 77–78.

11. Elaine Sciolino, "It's a Sea! It's a Lake! No. It's a Pool of Oil," *New York Times*, June 21, 1998.

12. Michael R. Gordon, "Russia to Offer U.S. Deal to End Iran Nuclear Aid," *New York Times*, March 17, 1999, 12.

13. Kelly Couturier, "Turkey Aims to Satisfy Its Fuel Needs," *Washington Post*, October 20, 1997, A17, A18.

14. "Khatami Calls for Expansion of Iran-Georgia Cooperation," *Xinhua News Agency*, June 14, 1998.

15. "Cooperation Agreement Signed at End of Georgian Foreign Minister's Visit," *BBC Summary of World Broadcasts/Vision of the Islamic Republic of Iran Network 1*, June 23, 1998.

16. Ahmed Rashid, *The Turkmenistan-Afghanistan-Pakistan Pipeline* (Washington, D.C.: Petroleum Finance Company, September 1997), 10.

17. For a more detailed discussion of Afghanistan's role in Caspian energy development, see, Shireen Hunter, "The Afgahn Civil War: Implications for Central Asian Stability," chapter 10 in this volume.

18. For a more detailed analysis of these trends and of the expectations for energy revenues in Turkmenistan, see Nancy Lubin, "Turkmenistan's Energy: A Source of Wealth or Instability?" chapter 6 in this volume.

19. Anna Kurbanova and Lyudmila Glazovskaya, "Turkmen-Iranian Relations Important Factor in World Policy," *TASS News Agency*, July 8, 1998.

20. "Turkmenistan, Iran Sign Joint Communique, Agreements," *BBC Summary of World Broadcasts/IRNA*, July 11, 1998.

21. "Turkmenistan, Iran Sign Joint Communique, Agreements," *BBC Summary of World Broadcasts/IRNA*, July 11, 1998.

22. "Iranian Anger over Turkmen Pipeline Deal," *RFE/RL Iran Report 2*, no. 9 (March 1, 1999): 3.

23. "Survey of Central Asia," *The Economist*, February 7, 1996: 12.

24. "Kazakhstan to Develop Ties with Iran Despite US Objections," *BBC Summary of World Broadcasts/TASS News Agency*, November 15, 1997.

25. "Irani, Kazakh Firms Discuss Oil Cooperation," *Xinhua News Agency*, March 14, 1998.

26. In the spring of 1995, Iran signed a surprise deal with the American company

Conoco to develop off-shore gas in the Persian Gulf at South Pars. Conoco beat out the French company, Total, for the contract. Under pressure from the U.S. Congress, the Clinton administration requested that Conoco's parent company, Dupont, drop the deal. In September 1997, three foreign companies—Total (France), Gazprom (Russia), and Petroneas (Malaysia)—signed a $2 billion contract with Iran to develop South Pars.

27. Buy back arrangements are structured such that foreign companies finance the up-front costs of energy projects and receive payment in output once the projects are on stream. Payment includes recovery of the investment plus an agreed profit. The scheme is not particularly popular with investors since they have to take all the risks and, under the present arrangements, cannot show crude oil or gas reserves as part of company assets. Most foreign companies prefer production sharing agreements (PSAs) but these are presently banned in Iran. See *Middle East Economic Digest*, July 16 1999: 15.

28. See the four part series in the *Financial Times*, July 1, 1998, including: Robert Corzine and Robin Allen, "Deals Shaped with an Eye to Constitution," 4; Robert Corzine, "Tehran Reopens Doors to Case Oil and Gas Resources," 4; Robert Corzine, "Top Companies Lured by Huge Opportunity," and David Gardner, "U.S. Action Underlines Investment Needs," 4.

29. Eliyahu Kanovsky, *Iran's Economic Morass: Mismanagement and Decline Under the Islamic Republic* (Washington, D.C.: Washington Institute for Near East Policy, 1997), Appendix 4.

30. Elaine Sciolino, "Iran's Difficulties Lead Some in U.S. To Doubt Threat," *New York Times*, July 5, 1994, A1.

31. "Iran: Economic Structure," and "Iran: Quarterly Indicators of Economic Activity," *Economist Intelligence Unit Country Reports*, November 17, 1996.

32. Remarks of U.S. Secretary of State, Madeleine K. Albright, Asia Society Dinner, Waldorf-Astoria Hotel, New York, June 17, 1998.

33. Federico Peña, Secretary of Energy, statement before the Committee on International Relations, U.S. House of Representatives, April 30, 1998.

34. Federico Peña, Secretary of Energy, statement before the Committee on International Relations, U.S. House of Representatives, April 30, 1998.

35. The Center for International Political Economy and the James A. Baker III Institute for Public Policy, *Unlocking the Assets: Energy and the Future of Central Asia and the Caucasus* (Houston: Rice University, April 1998).

36. "Countries Delay Caspian Sea Agreement, IISS Questions Oil Potential of Region," *The Energy Report 26*, no. 18 (May 4, 1998).

37. See Martha Brill Olcott, "The Caspian's False Promise," *Foreign Policy*, Summer 1998: 95–113.

38. David Filipov, "U.S. Warns against Pipeline Going through Iran," *Boston Globe*, March 1, 1998, A12.

9

Paradigms for Russian Policy in the Caspian Region

Peter Rutland

Russian policy toward the development of the energy resources of the Caspian Basin is a complex subject for analysis because it nests within several broader sets of policy concerns: relations with the United States, which is also actively pursuing its own policy in the region; relations with the countries of the "near abroad," the term Russians use to refer to the other former Soviet states; and policy toward Russia's own domestic energy sector.

Figuring out exactly what role energy factors have played in Russian foreign policy since 1991 is no easy task. Oil is both an end in itself (for the wealth it generates) and a means to an end (aiding the projection of Russian power and influence). Energy interests have pulled Russian policy in different directions at different times, and often in different directions at the same time. No single model can capture these conflicting and confusing pressures and trends. Studying the role of oil in Russian foreign policy is like trying to assemble a jigsaw puzzle where the pieces from five different puzzles have been mixed together, and where one has no picture of what the completed puzzle is supposed to look like.

Some analysts posit that policy is driven by a conspiracy of elite interests, others that there is no discernible policy at all. Some of these paradigms rely on universal economic theories, and some on the logic of geopolitics. Few of the authors advancing these competing paradigms spend much time on the economics of energy per se.

Above all, it must be remembered that Russia is not pursuing an integrated, consistent policy in its relations to Caspian energy development. Policy conception and execution are divided among a variety of domestic political and economic agencies, and there is no single body capable of coordinating these various efforts. While the "rational actor" model may have been of some use in examining the Soviet Union's foreign policy, it would be completely misleading to look for an integrated conception of Russian national interests behind its foreign policy toward the Caspian since 1991. The Ministry of Foreign Affairs has proved singularly incapable of playing such a coordinating role. It is particularly weak with regard to the coordination of foreign economic relations (which loom large in policy toward the Caspian region). The Security Council has been a somewhat more effective organ, although focusing more on military and nuclear affairs than on diplomatic or economic relations.

The complexity of the internal structure of Russian foreign policy is more than matched by the complexity of the Caspian Basin itself, which presents a complicated and evolving set of political and economic conundrums. Even if Russia spoke with one voice in its policy toward the region, it would still not be able to come up with a coherent set of policies. There are simply too many unknowns, from the uncertainties of political succession of aging leaders, to the possibility of a shift in U.S. policy toward Iran. Even the overall volume of oil and gas actually to be found beneath the Caspian Sea is still uncertain.

COMPETING PARADIGMS

What patterns emerge from this complex web of geopolitical commitments and contentions? Most discussion of the energy–foreign policy nexus in Russia can be grouped into one of the following schools of thought. Some observers believe Russian policy is responding to the logic of an unfolding economic model: perhaps energy-dependency; perhaps liberalization; possibly a looting of the national treasury by a narrow group. Others see energy policy as a by-product of deep geopolitical forces that are ineluctably shaping Russian foreign policy. Alternatively, policy may be tossed from one side to another in response to pressure from competing groups, none of them capable of providing it with a clear direction.

1) The Kuwaitization Model

According to this school, Russia's vast endowment in natural resources means that its comparative advantage in the international division of labor lies in the exploitation of its energy fields. The boom in Russian exports since 1993 has been driven by sales of oil and gas, and by sales of other minerals such as steel and aluminum whose processing is dependent on cheap energy. Energy accounts for half of Russia's export earnings, a third of federal budget revenues, and a quarter of the overall Russian GDP.

Proponents of "Kuwaitization" argue that an energy boom is the best hope (perhaps the only hope) for Russia's economic renaissance, especially if contracts for pipe and machinery are placed with Russian companies rather than western suppliers. Opponents, whose ranks include Moscow Mayor Yurii Luzhkov, argue that such a strategy will condemn Russia to subordinate status as a "raw materials appendage" of western multinational corporations, who will deny Russia a fair share of the proceeds and who will not allow Russian engineering plants to pick up supply contracts.[1] For the patriotic opponents of Kuwaitization, production-sharing agreements (PSAs) are an infringement on Russian sovereignty, and something which countries like Iran and Saudi Arabia have rejected since the 1970s. Hence the State Duma's glacially slow adoption of PSA legislation and subsequent project authorization, in defense of Russia's "economic security." A law on production-sharing was adopted in November 1995, but required that each project be subject to parliamentary approval. It was not until July 1997 that the Duma approved the first batch of seven sites for development by foreign multinationals—instead of the hundreds proposed by the government. By December 1998, only two projects had actually broken ground (Sakhalin 1 and 2), at which point the Duma relented and amended the law, easing the restrictions on PSA agreements.[2] This should clear the way for the approval of more sites, with possibly up to 30 percent of Russia's new oil deposits becoming open to foreign investors. Although the Duma did start approving more projects, troubling limits remain (for example, an insistence that 70 percent of the equipment used must be bought from Russian manufacturers).

The Kuwaitization model implies that Russia is a passive victim of international economic forces. However, foreign investors have in fact been very slow to penetrate the domestic energy industry. Instead, Russian corporations have emerged as powerful forces, subsidizing regional authorities and providing some 40 percent of federal government revenues.[3] The only major example of foreign involvement since the PSA law was amended has been British Petroleum's investment of $570 million in return for a 10 percent stake in Sidanko in 1997. BP's bold move in acquiring a stake in Sidanko backfired in the wake of the August 1998 financial crisis, when Sidanko plunged into insolvency with debts of $460 million. In 1999, BP found itself engaged in a fierce legal battle with Sidanko's domestic creditors, who were seeking control of parts of the company—most notably the Tyumen Oil Company, which acquired Sidanko subsidiary Kondpetroleum in October and Chernogorneft in November. As BP Amoco mounted an international campaign of protest, Tyumen Oil relented, returning Chernogorneft to Sidanko in December.[4] But the whole experience was a bruising warning to potential foreign investors of the perils of the Russian market.

Accordingly, this paradigm, shorn of its negative political rhetoric, in fact implies that Russia's oil and gas corporations will play a dominant role in determining Russia's general policy toward the near abroad, seeking to keep out potential competition from Caspian Sea fields and/or to maneuver themselves into a share in those projects. If energy interests are playing a dominant role in shaping Russian foreign policy, then one can see that it can be pulled in contradictory directions. On one side, the Caspian Basin resources are competing with Russian-sourced materials; on the other side, if they are going to be brought to market anyway, Russian companies will want to participate in the projects.

Thus even if "Kuwaitization" is indeed the dominant force shaping Russian economic development, this will not provide us with easy answers to the course of Russian policy toward the Caspian.

2) Market Forces

For the liberalization school, economic events in Russia are being driven by the ineluctable tide of the global market economy, unleashed by the liberalization program begun by Prime Minister Yegor Gaidar in 1992. Any efforts by the state to "pick winners" and run an active industrial policy would merely allow the old vested interests of the command economy to creep in through the back door and would be a waste of resources. The government was to focus on the mantra of liberalization, stabilization, and privatization, and success in these measures would be in the general interest of Russian society. The policy was delayed and distorted by opposition from entrenched interests but supported by international financial institutions, who often made new loans conditional on progress in these three areas.

Despite being unpopular with the general population, and having few discernible bases of support within political and economic elites, the Yeltsin-backed governments of Gaidar and, later, First Deputy Prime Minister Anatolii Chubais were indeed able to ram through most of the liberal agenda from 1992 to 1995.[5] The energy sector had escaped much of the onslaught of price liberalization and privatization in 1992. But 1995 saw the removal of oil export quotas and the introduction of the loans-for-shares scheme, which was a way of transferring some of the major oil companies into private ownership in the absence of Duma legislation authorizing the second wave of cash privatization.

Even with the extensive liberalization, there is still plenty of scope for companies to bargain with government agencies. Although oil export quotas were formally abolished at the IMF's insistence in 1996, companies were still only allowed to export an average of 27 percent of their oil production, and the criteria for awarding access to pipelines remained rather opaque (with, for example, Rosneft allowed to export 52 percent of its oil).[6] Tax arrears were another bone of contention. Oil and gas enterprises owed some 19 trillion rubles to the federal budget in late 1997—but in turn were owed more than 100 trillion rubles by their customers, leaving plenty of scope for bargaining over tax offsets.[7]

The defenders of liberalization argue that market reform offers the best hope—the only hope—of economic revival in Russia. They note that Russian consumers seem to prefer the full shelves (even with thinner wallets) of the market economy to the economics of shortage that prevailed prior to 1992. Ordinary Russians relish the opportunity to purchase imported goods, and the freedom to change rubles into dollars. At the international level, liberalization has won Russia membership (sort of) in the exclusive G7 club and the prospect of entry into the World Trade Organization. The considerable progress toward liberal market policies implies that, contrary to the Kuwaitization model, Russian policy has not been captured by a single powerful group of interests. The market forces paradigm would imply that Russian companies can be persuaded to take a constructive role as partners in the development of the Caspian Basin's resources and that one can expect them, and not military interests, to be the driving forces behind Russia's policy toward the region.

3) Rent Seeking

A cynical position is to argue that the Russian oil and gas industry is a classic example of the triumph of rent seeking over profit seeking.[8] A narrow circle of elites was fortuitously positioned by the Soviet collapse to loot Russia's energy resources, and they will continue to do so until the cookie jar is empty. The oil and gas industry had been built up under central planning, with decades of costly investment. The Soviet collapse shattered the planning system which channeled resources into the sector, but left the infrastructure intact—and left a small number of managerial and political elites with the opportunity to divert the proceeds from oil and gas exports without having to worry about ensuring the renewal of the production infrastructure.

Rent seeking involves the exploitation of a monopolistic position and is contrasted with profit seeking, which involves the pursuit of wealth in a competitive market environment. The Russian rentier elite enjoys a monopoly in terms of their political access to the oil industry infrastructure and their consequent ability to skim revenues from oil and gas exports. Rent-seeking elites would adopt a short-term, disruptive posture toward Caspian development, seeking to block or delay rival production coming on stream in order to preserve their grip on the Russian industry for as long as possible.

The rent-seeking model implies that the current situation is transitory. Sooner or later the infrastructure will decay, and production will fall below the point at which current levels of domestic consumption and exports can be sustained. Unless domestic consumers can be forced to accept a lower supply level, the available rents will start to evaporate. At that juncture the elite will (or at least should) shift over to a nationally managed policy for development of the energy sector (Kuwaitization) or will be obliged to open up to market forces, including access for new owners. According to this paradigm, however short-sighted and predatory Russian energy policy has been over the past decade, western policymakers should adopt a more patient, long-term view in the expectation that Russian policy will shift toward a more cooperative and market-responsive direction.

4) "Russian Bear"

Some argue that behind the apparent chaos of governmental decision making, one can discern the inexorable logic of national interests.[9] Russia is still a great power, with geopolitical interests stretching at least across the near abroad. However, Russia's military forces are in disarray, and, moreover, the international community will not allow Russia to use military force to rebuild its sphere of influence. If Russia can no longer rely on military means to project its power, it must thus turn to economic levers of control. Hence the widely cited May 1996 theses on Russian policy toward the near abroad, published by the nongovernmental Moscow-based Council on Foreign and Defense Policy, which urged that Russia avoid military methods and should instead use its economic muscle to become the "natural" center of the Commonwealth of Independent States (CIS).[10]

In this context, Russia's energy corporations are seen as a surrogate for the country's deep-rooted national interests. Russia is energy-rich, while its western neighbors are energy-poor and its eastern neighbors are dependent on Russian goodwill for the easiest access to international markets. Russia is seen as exploiting the energy debts of countries such as Ukraine, Belarus, and Moldova to extract political concessions and perhaps acquire equity shares in their energy industries. In response, the newly independent countries engage in efforts to develop alternative energy supplies (such as oil and gas from Central Asia) and to find alternative routes to world markets, which are seen as vital to ensuring the real sovereignty of the newly independent states. Some commentators coming out of this geopolitical perspective point to the emergence in 1997 of a GUUAM axis (Georgia-Ukraine-Azerbaijan-Moldova, now extended to include Uzbekistan), uniting the southern tier of countries in their mutual determination to check Moscow's erratic policies.[11] According to this line of thinking, the strengthening of organizations such as GUUAM, which exclude Russia, and the increased U.S. presence in the southern tier of newly independent states, will only serve to deepen Russian suspicions. Rather than promoting a spirit of cooperation, such U.S. policies may actually stimulate the opposite of what they intend—that is, a hardening of Russian policy in the region.

5) Pluralism

A fifth approach is to view Russian policymaking as an open-ended process in which rival groups compete to influence policies as they flow through the decision-making procedure. In this "garbage can" paradigm, the dominant picture is one of fragmentation and confusion, since the tightly integrated decision-making process of the Soviet era has given way to a chaotic free-for-all.

With the dissolution of central authority, no clear decision-making procedure can be discerned at all.[12] There is no transparent division of power between the Parliament and the executive branch. Sometimes the executive branch chooses to implement policy by decree and ignores the legislature. At other times, the government defers to the State Duma and seeks a legislative framework for its actions. The policies emanating from the governmental machinery are often at odds with decrees coming out of the presidential apparatus. Government policy itself can be rather schizophrenic, with the various deputy prime ministers (at times as many as eight!) espousing a broad range of agendas. New agencies such as the Security Council and the Defense Council have been created, but have had little success in imposing order on the situation. Foreign policy toward the CIS often seems to be driven by independent actors on the ground—from military commanders in the Caucasus and trans-Dniester to energy corporations like Gazprom and LUKoil. (For example, nobody has owned up to responsibility for the 1994 transfer of $1 billion worth of Russian weapons to Armenia.) Adding to the confusion has been the decentralization of power from Moscow to the regions. The newly empowered presidents and governors of Russia's regions have considerable leeway to try to run their own foreign policy, from attracting foreign investors to managing relations with neighboring states.[13]

This all amounts to pluralism run amok, where the "decision-making process" seems to be reinvented for each decision. In order to understand policy, whether it be the decision to build the Primorsk oil terminal on the Baltic coast, or the plan to build a new pipeline around Chechnya, one must construct a separate model for each issue, listing the interests for and against *and* the institutions through which they try to operate.

THE POLICY PROCESS

There do seem to be some regularities in the policy process. For example, the foreign ministry seems able to exercise more control over policy toward the "far abroad" than toward the CIS. Russian policy toward China, Iraqi sanctions, and NATO expansion is clearly espoused and skillfully pursued, and correspondingly seems to enjoy a fairly high degree of consensus among Russian elites and (to the extent that they care) the Russian public at large. The same could not be said for virtually any aspect of policy toward the CIS. This is presumably a reflection of the historical continuity of the foreign-policy-making apparatus vis-à-vis the outside world and of the need to create institutions (and policies) from scratch to deal with the CIS. Foreign policy toward the "far abroad" becomes somewhat more contentious when one moves from the political to the economic arena. Issues such as IMF lending or access for foreign investors generate considerable controversy.

The government has had some successes in enforcing a single policy toward the CIS—in stubbornly pushing ahead with trade liberalization and the introduction of the "ruble corridor" in the summer 1993, for example. This, it should be noted, occurred during a period when Yeltsin was in the ascendant and heading for a showdown with the Parliament. Since then, some of the fiercest "under the carpet" arguments between rival factions of Russia's ruling elite have taken place over policy toward the near abroad. They are divided over whether to make economic concessions in return for political support and economic cooperation. The energy lobby generally favored cutting back on deliveries to CIS countries unable to pay for them, since the products could profitably be switched to western markets. However, people like former First Deputy Prime Minister Oleg Soskovets (1994–96), a product of the metallurgy lobby, believed that it was important to preserve Russia's market share in the CIS, even if it meant selling energy to them at less than world prices (and selling to customers who could not pay). This was both for political reasons and because many Russian manufacturing plants (aluminum smelters, for example) were dependent on supplies from Ukraine and Kazakhstan (and vice versa). Overall, however, "with regard to the CIS, Russia seems to be in a state of confusion and helplessness."[14] It has been singularly unsuccessful in mobilizing support from CIS countries in its diplomatic initiatives—such as in its failure to prevent NATO expansion.

Clearly, the energy corporations play a large if not dominant role in shaping Russian foreign policy toward the near abroad. As an outsider, one can only guess exactly how this works in practice. It does not look as if there is any organized, structured process through which policy is formulated. Rather, it seems to be driven by the same process of informal bargaining that has come to dominate most of Russian domestic policymaking. It is hard to say where domestic politics ends and foreign policy begins. The deals and favors that oil companies trade with government officials with respect to foreign projects are seamlessly connected to the same process in the domestic arena. Bargaining ranges over a broad gamut of activities: from sweetheart privatization deals to export pipeline quotas; from tax arrears waivers to "goods credits" (*tovarnye kredity*) for farmers. The bargaining spills over into the political realm: the same corporations bankroll political campaigns and fund newspapers that energetically engage in mud-slinging against rival politicians.

Several general rules of the game seem to have emerged in recent years. First, there are strong elements of continuity. The vested interests resist efforts to disrupt the status quo and have to be given something in order to induce them to accept a policy change. Note for example Gazprom's success in beating off the reformist challenge from First Deputy Prime Minister Boris Nemtsov, who took office in March 1997 with a pledge to break up the utility monopolies and introduce competitive pricing. Gazprom President Rem Viakhirev was explicit in rejecting Nemtsov's suggestions for political, not economic, reasons: "We are not simply producing and selling gas. We are pursuing a popular, socially oriented policy whose significance goes beyond narrow economic calculation."[15]

Second, the process of privatizing the remaining state-owned oil corporations, both in the 1995 loans-for-shares scheme and in the 1997 cash auctions, showed that the good old Soviet principle of "an earring for each sister" still held sway. That is, favors and special projects were doled out in turn to the rival consortia. The feeling that this understanding was being violated—in favor of Vladimir Potanin's Oneksimbank (which controls Sidanko)—triggered the bankers' attack on Chubais in the summer of 1997.

Third, as the market economy slowly sinks firmer roots in Russia, one can expect that the autonomy of these companies vis-à-vis the government will gradually increase. The ability of these companies to access world capital markets means they will eventually grow less dependent on financial support from the Russian government (in the form of allotted shares in World Bank financing, for example).[16]

The consolidation of powerful, independent energy companies in Russia should make it less easy for conservative forces in the Russian government to manipulate Caspian policy in the interests of national security and should give the United States more opportunities to find partners with an interest in a cooperative development strategy.

ASSESSING RUSSIAN POLICY

Each of these radically differing perspectives on Russian foreign and economic policy can be found in western writing on Moscow's policy toward the Caspian countries. Each perspective can find evidence and examples to support its interpretation. But it would be unwise to declare that any of these interpretations is clearly superior or inferior to any of the others. Russian policy has been chaotic and contradictory, and there are too many unresolved questions with respect to the future trajectory of developments in the Caspian region to come up with a definitive assessment of Russian foreign policy.

Overall, Russia's foreign policy toward the near abroad has been opportunistic rather than strategic and is negative (blocking initiatives, stalling progress) rather than positive (creating new openings and developing new projects). Russia is not the source of the poverty and strife that has gripped much of the newly independent countries, but neither is it doing much to help solve the situation. Though weakened, Russia still has a broad variety of legal, political, military, and economic tools that it can and does deploy to advance its interests. The Russian energy corporations seem to play a dominant role in formulating and actually implementing Russian foreign policy in many crucial respects, although the Russian military cannot be written off as a player (witness its return from defeat in Chechnya in 1996 to something approaching victory at the beginning of 2000).

In this situation one can see the *potential* for the emergence of a neoimperialist strategy that suits both the oil companies and the "power ministries," who claim to speak for Russia's national security interests. In this scenario, the Russian energy corporations and associated financial groups will move into the countries of the near abroad and gain political influence using the same tactics of financial manipulation that they have been perfecting inside Russia itself. Some countries are more vulnerable to this tactic than others. Such a neoimperialist strategy would receive broad support across the political spectrum in Russia.

The scope for neoimperialist ventures has been somewhat eroded by developments over the past two years, although the Caspian Basin remains highly unstable. First, the defeat of the Russian army in Chechnya in August 1996 was a powerful blow to the prestige and confidence of the Russian military establishment. Most commentators assumed that the military would not be in a position to mount a second invasion of Chechnya, still less pursue such an attack to a successful conclusion. However, Chechen raids on neighboring Dagestan in August 1999 did trigger such an invasion, and newly-appointed Prime Minister Vladimir Putin gave the military a free hand to pursue the war to a conclusion using whatever means military leaders deemed necessary. Russian victory over Chechnya may encourage them to increase their military pressure on Georgia and perhaps Azerbaijan.

Second, a countervailing trend has been the steady increase of U.S. and NATO influence in the region. In this regard, a symbolic corner was turned with the Partnership for Peace exercises that took place in Central Asia, with the involvement of U.S. troops, in November 1997.

Third, the slump in the world oil price in 1997 and 1998 from $20 to less than $10 a barrel undercut the profitability of Russian oil and gas companies, weakening their bargaining position and driving them closer to the Russian government for assistance. However, the rise in the world oil price to more than $30 a barrel, which followed OPEC's March 1999 decision to restrict output, has restored their profitability. On the other hand, the oil price rise also had the effect of making the energy resources of the Caspian suddenly look more attractive to western investors, many of whom had been developing cold feet given the slow pace of development in the region.

Fourth, the financial collapse of August 1998 emptied the coffers of the leading Russian financial oligarchs and weakened their room for maneuver. The impact of the financial crisis on the energy barons, in contrast, was generally favorable. The radical devaluation of the ruble (from 6 to 28 per dollar) has done much to restore their profitability by lowering their domestic costs, while the crisis weakened the financial oligarchs who were looking to win ownership over still more oil assets.

Looking back at the trajectory of Russian policy toward the newly independent states since 1991, it is remarkable to what extent political and military considerations have taken a back seat to economic concerns. For example, the much-feared presence of a 20-million strong Russian diaspora in the near abroad has had a negligible impact on Russian foreign policy, with the partial exception of relations with Estonia and Latvia. The division of the military assets of the Soviet army went more smoothly than anyone would have expected—most notably, the withdrawal of nuclear weapons from non-Russian territory. The lingering dispute over the division of the Black Sea, eventually settled in 1997, was more an exercise in historical symbolism than a matter vital to the security concerns of either Russia or Ukraine.

There are those who argue that Russia's desired shift from military to economic dominance may not work, since the Russian economy is too weak to play such a role.[17] Also it is important to remember that the Russian oil corporations cannot do it on their own: they will need western capital and technology to develop the new fields and pipelines that the region needs.[18]

From the outside, Russian policy appears to be driven by institutional rivalries between military interests and energy companies (which are themselves split into competing teams). On the other hand, Yurii Fedorov suggests that since 1996 a consensus has developed over the need to consistently advance Russian interests in the region—a consensus forged largely in opposition to U.S. efforts to penetrate the area.[19] Thus Russian policy toward the Caspian Basin is driven by contradictory pressures—to cooperate with and to oppose U.S. companies moving into the region.

THE CHALLENGE OF ECONOMIC DEVELOPMENT IN THE CASPIAN BASIN

In order to understand Russian policies toward the Caspian Basin, it is important to establish the scope and complexity of the challenges facing the region. The Caspian littoral states are blessed with vast reserves of oil and natural gas. Some of the earlier rhetoric about an energy cornucopia has ebbed: the region probably holds about 5 percent of the world's oil and gas reserves, while Russia itself holds 20 percent.[20] But there is certainly enough oil and gas in the Caspian Basin to serve as an engine of development for the region's economies, and the region's leaders have made the attraction of foreign investment the centerpiece of their strategy since attaining independence in 1991. Kazakhstan had received $3.2 billion in energy-related investment by 1998, while Azerbaijan gained $1.8 billion.[21]

However, eight years later, the states of Central Asia and the South Caucasus remain mired in poverty. In Kazakhstan, oil production fell by one quarter before recovering to roughly the 1991 level, while Azerbaijan in 1999 was still exporting only some three million tons a year (from total output of eleven million), 10 percent below the 1991 level. Progress has been hampered by a broad range of factors: lack of market institutions and basic infrastructure, the slow pace of privatization, and corruption and bureaucratization.

The main challenge facing the region is geographical isolation. The countries are landlocked and at least 2,000 kilometers (km) from ports on the open seas: hence getting their resources to world markets will be difficult. (Access from the Black Sea to the Mediterranean and beyond is restricted by Turkish unease at increasing shipments through the Bosporus Strait.) The second major task facing these countries is to achieve a degree of political stability and resolve ethnic conflicts. The third challenge is to establish a stable pattern of regional cooperation: to turn Russia from a threat to a partner and to solve the ongoing dispute over the legal status of the Caspian Sea. The U.S. government believes that the solution to each of these problems is one and the same. Exporting resources will create the wealth that will facilitate long-run political stability and promote regional cooperation.

1) The Challenge of Getting Oil and Gas to Market

To date, most Central Asian oil and gas has been exported via Russian pipelines.[22] Oil extracted in western Kazakhstan is processed in Russian refineries: Kazakhstan's own refineries in the east of the country import oil for domestic use from Siberia. Similarly, the country exports natural gas from its western Karachaganak field to Russia, but must import gas from Uzbekistan and Turkmenistan for its eastern cities.

The Clinton administration argues that it is good to engage Russia as a partner in the development of the region, but that these countries must lessen their dependence on Russian export routes. Since 1991 Russia has repeatedly proved its willingness to exploit its monopoly position as controller of the existing export pipelines to push down the price of Kazakhstani oil or Turkmen gas—or to shut down exports completely, as it did to Turkmenistan in 1994. Russia subsequently allowed Turkmen gas to supply Ukraine (but not to customers further west). Ukraine predictably fell behind in paying for gas deliveries, and Ashgabat cut off its supplies to Ukraine in March 1997.[23] In 1999, Russia increased pipeline quotas for Kazakhstani and Azerbaijani oil (from 6.4 million tons in 1998 to 11 million in 1999) in a bid to persuade them that it was unnecessary to develop an alternate export infrastructure.[24]

Russia has threatened (and implemented) cuts in access to export pipelines for Kazakhstan in order to win shares in the major energy development projects inside the country. In 1995, this tactic won Russia's Gazprom a 15 percent share in the Karachaganak field, which was being developed by British Gas and Italy's Agip. Russian obstructionism helped cause Kazakhstani gas output to slide from 4.2 billion cubic meters in 1991 to 2.5 billion in 1995. Russian policy toward Kazakhstan is mediated, if not actually driven, by its major energy corporations: Sidanko and LUKoil compete to supply Kazakhstan's Pavlodar refinery, and the state-owned Transneft company, which owns the Russian oil pipeline network, negotiates quotas for the pipeline access.

Kazakhstan's main hope for revival has been the giant Tengiz field, on which Chevron has now spent $1 billion through the joint extraction venture Tengizchevroil. Chevron holds 45 percent of the venture, the Kazakh partner Tengizmunaigaz 25 percent, and Mobil 25 percent. In January 1997, LUKoil bought a 5 percent stake in the consortium for $200 million. In 1992, the international Caspian Pipeline Consortium (CPC) was formed by Kazakhstan, Russia, and Oman to build a new 1,500 km, $2 billion pipeline from Tengiz to Astrakhan in Russia and on to the Russian Black Sea port of Novorossiisk, with a projected capacity of 70 million tons a year.[25] Half the CPC shares are held by governments and half by companies. Chevron subsequently found the economics of the pipeline unattractive, and work was delayed. After a restructuring in 1996, Chevron's share was cut. At present the governments of Russia, Kazakhstan, and Oman hold 24 percent, 19 percent, and 7 percent, respectively. With regard to the private companies, Chevron holds 15 percent; LUKArco (a joint venture between LUKoil and ARCO) 12.5 percent; Rosneft-Shell Caspian Ventures 7.5 percent; Mobil 7.5 percent; Italy's Agip 2 percent; and three smaller companies the remaining 5.5 percent. Actual laying of pipe finally began in November 1999, with initial volume of 28 million tons of oil per year expected.

Kazakhstan produced 26 million tons of oil in 1997, of which 7 million were exported through Russia, 1 million through Baku, and 1 million in a swap deal with Iran (delivered across the border in railcars).[26] Yet it is likely to be a number of years before Kazakhstan's dream of oil wealth is realized.[27]

Azerbaijan has been able to pursue a more independent line than Kazakhstan, although after the fall of the nationalist government of Abulfaz Elchibey in 1993, steps were taken to bring in more Russian partners. Under President Heydar Aliyev, the government has tried to involve companies from as many countries as possible in order to raise Azerbaijan's diplomatic profile, with a view to increasing the chances for resolution of the conflict with Armenia over Nagorno-Karabakh.[28]

The Caspian Sea has some 4 billion tons of proven oil reserves, and the Azerbaijan International Operating Company (AIOC) should be producing 35 million tons a year by 2010. However, as John Maynard Keynes noted, there is a difference between the short run and the long run ("in the long run, we are dead"). Due to the disruption following the Soviet collapse, Azerbaijani oil output actually fell from 12 million tons in 1991 to 9 million in 1996, recovering to 11 million in 1998.[29] Only in October 1997 did the first oil from the offshore Chirag field start to come ashore at Baku, and this will constitute a mere 5 million tons for the next couple of years.

One factor holding back production has been an acute shortage of drilling rigs. New heavy equipment can only be brought in through Russia's Volga-Don canal system, and Russia has been denying passage to many Azerbaijan-bound vessels. By the end of 1999, however, the AIOC had produced enough oil to pay off the development costs and was starting to yield a profit.[30] In June 1999, BP Amoco announced the discovery of large offshore natural gas fields in Shah Deniz (containing up to 400 billion cubic meters). Gas, rather than oil, may turn out to be the key to Azerbaijan's future, but this may well cause a renegotiation of plans to build a trans-Caucasus oil export pipeline from Azerbaijan and a gas export pipeline from Turkmenistan.

2) The Case of Chechnya

The interdependence of oil and politics is vividly illustrated by the case of Chechnya. Many—including former Security Council Secretary Aleksandr Lebed—believe that the Chechen war was started in December 1994 by financial interests keen to get a piece of the oil action bubbling up in Azerbaijan. The timing seems more than coincidental. For three years, Moscow had effectively ignored Chechnya's November 1991 declaration of independence, but then sent Russian forces into Chechnya just two months after the signing of the "contract of the century" with the AIOC in Baku. Chechnya has its own modest oil field and before the war had a large refinery (with a 19 million ton annual capacity). More importantly, it sits astride what at that time was the only export pipeline from Azerbaijan. Somehow or other, the trans-Chechen pipeline continued to operate through the first year of the 1994–96 war, although the line was tapped in literally hundreds of places in order to provide fuel for back-yard oil refineries. The line appears to have ceased operations sometime in 1996. The signing of the Khasavyurt treaty between Russia and Chechnya in August 1996 paved the way for the withdrawal of Russian forces by January 1997, the signing of a peace treaty in May, and the establishment of de facto Chechen independence.

Throughout 1997 sporadic negotiations continued between Grozny and Moscow, with the latter holding out the carrot of economic cooperation but the former insisting on Russian recognition of Chechen independence. Boris Berezovskii was brought into the government as deputy secretary of the Security Council, ostensibly to forge a business deal that would lure Chechnya into agreeing to some sort of status within the Russian Federation. Berezovskii's deal failed to materialize, although the pipeline itself was rehabilitated and oil began flowing again in December of that year. However, exasperated by the continuing series of terrorist incidents, Moscow proposed building a new $200 million, 250 km oil pipeline bypassing Chechnya. (Similarly, in 1997 a new 80-km railway was built linking Dagestan with Rostov without crossing Chechen territory.) It was not until November 1999, however, that work on the by-pass pipeline apparently began.[31]

The Kremlin insists on keeping Chechnya within the Russian Federation, and sees the transshipment of oil across the republic as a way to help rebuild the republic's shattered economy.[32] Russia's policy toward Chechnya has been driven primarily by national security concerns, and, with the outbreak of the second Chechen war, responsibility for running the policy has been handed almost entirely to the defense and interior ministries. There were efforts to use the carrot of economic interests (from oil projects to reconstruction aid) to persuade the region's leaders to rejoin the Russian Federation between 1994 and 1996, and again in April 2000, but these steps failed dismally (while providing ample opportunities for corruption).

Russia's problems with separatist violence are not confined to Chechnya. Both the Chechen and northern pipelines transit Dagestan, where there have been frequent clashes between the Chechen minority living in the republic (Akins) and other ethnic groups. The one-million strong Lezgin minority lives astride the Russian-Azeri border and also has an incipient separatist movement. (Lezgin activists were charged with detonating a bomb in the Baku subway in 1994.)

3) The Caspian Dispute

The third challenge for economic development in the Caspian is to resolve the serious dispute over the legal status of the Caspian Sea itself. After the dissolution of the Soviet Union, Moscow and Tehran originally stuck by the positions fixed in treaties signed in 1921 and 1940, which treated the sea as a lake from the point of view of international law. This meant that the exploitation of its resources beyond a coastal strip must be the joint endeavor of all the littoral states. Azerbaijan and Turkmenistan have argued that the sea should be divided up among the littoral states by extending the sovereign territory of each country outwards until it meets the others in the middle, as under the 1982 UN Convention on the Law of the Sea. This solution would give Azerbaijan and Turkmenistan exclusive rights to the most promising offshore fields. Kazakhstan has wavered between the two positions, eventually gravitating toward the Russian interpretation.

In subsequent years, however, Turkmenistan shifted from its initial pro-Azerbaijani stance to an ambiguous position, and in July 1997 challenged Azerbaijani plans to develop the disputed Kyapaz field. (Partly the dispute revolves around such arcane details as how to treat off-shore islands and irregularities in the coastline, when calculating the median line between the coasts of Azerbaijan and Turkmenistan.) Russia tried to compromise by expanding the allowed coastal strip to 75 km, but this did not resolve the disagreement between Azerbaijan and Turkmenistan, since most of Azerbaijan's off-shore fields, unlike those of Kazakhstan, lie more than 75 km offshore.

Russia started to shift its own position in response to Azerbaijan's willing-ness to grant Russian companies a share in oil development projects within its own zone. For example, LUKoil won a 33 percent stake in the Karabakh field and 10 percent stakes in the AIOC and in the consortium developing the Shah Deniz fields. Since April 1998, Russia has been agreeable to a national subdi-vision of the seabed, while still favoring joint management of the waters and fisheries. In part this shift was the result of a realization that contrary to prior expectations quite rich energy deposits probably lie in the northern, poten-tially Russian, zone. In July 1998, President Saparmurat Nazarbaev of Ka-zakhstan signed a treaty with Russian President Boris Yeltsin agreeing on the territorial division of the offshore seabed between Russia and Kazakhstan—an agreement that recognized the Russian-backed principle of joint develop-ment of the water column. Even if all five powers eventually agree to territo-rial division of the seabed, the task remains to agree on the demarcation of the zones between Azerbaijan, Kazakhstan, Turkmenistan, and Iran. Iran has been holding out for splitting the seabed into five equal portions—since the Iranian share based on coastline (under the international convention for seabed rights) would give it only 10 percent of the seabed.

4) Competing Export Routes

Washington advocates a dual-track approach to export pipelines: expand-ing exports through Russia while also building a new major export pipeline across the South Caucasus.[33] The policy was reaffirmed by U.S. Commerce Department Jan Kalicki at a Moscow conference on pipeline projects in Octo-ber 1999.[34] This approach is designed to give Russia a stake in the develop-ment of the Caspian, while at the same time giving the countries of the region an independent export route should relations with Russia turn sour at some point in the future. From the U.S. point of view, therefore, it kills two birds with one stone: it provides incentives for Russia to cooperate, while guaran-teeing the security of the newly independent states. From the Russian point of view, however, it keeps two birds alive. The Russian pipeline is a carrot for Russian doves, while the southern pipeline is a stick for Russian hawks, who see it as part of a sinister plot to project U.S. influence (with a military pres-ence sure to follow) in Russia's "backyard."

Apart from Russia, there are four alternative export routes for Caspian oil and gas—west across the Caucasus mountains to the Black Sea; south across Iran; east across China; and southeast across Afghanistan.[35] The western route across the southern Caucasus is the export corridor most favored by

Washington. There already exists a narrow-diameter pipeline from Baku to Supsa, on the Black Sea coast of Georgia, which has recently been rehabilitated. There have been intense and ongoing negotiations over a plan to build a major new oil export pipeline to Supsa. To gain access to the world market, oil would have to be shipped on from Supsa by tanker through the Bosporus Strait, something to which Turkey and the United States object on environmental grounds. In 1995 Turkey declared new limits on tankers passing through the Bosporus, arguing that the risk of ecological disaster was too high. Russia claimed that this was a violation of the 1936 Montreux international convention guaranteeing free passage through the Strait.

Instead of shipping oil through the Bosporus, Ankara wants to build a 1,800-km, $2.7-billion pipeline south from Supsa, across Anatolia, to Ceyhan on Turkey's Mediterranean coast. The proposed Supsa-Ceyhan pipeline is just as much a political as an economic project for President Aliyev of Azerbaijan. Aliyev hopes that involving outside powers such as Turkey and the United States in the economic development of his country will give him the diplomatic leverage he needs to bring about a resolution of the Nagorno-Karabakh conflict with Armenia—which remains the number one political challenge for Azerbaijan's president. The political attractiveness of the project (for Azerbaijan) is at least as important as its attractiveness on economic grounds.

As noted by Sabri Sayari and David Hoffman in this volume, the Turkish option is not problem-free, however. Instability throughout the South Caucasus and in the Kurdish-populated areas of eastern Turkey could endanger the project. Moreover, the commercial feasibility of such a pipeline across Turkey is uncertain, given the distances involved, the political risks, and questions about the actual amount of oil available in the Caspian region. The official U.S. position has been that the project must be viable on commercial grounds, and cannot be launched for its political advantages alone.[36]

To make the Turkish route commercially viable, oil may have to be brought from the fields on the eastern side of the Caspian, which would require the construction of an undersea trans-Caspian pipeline. This can only be done if the legal disputes over the status of the sea are resolved. Russian obduracy could hold up such a pipeline for many years, with the Russians using the argument that such a pipeline would threaten ecological catastrophe. The longer Russia can delay the creation of a new southern export route, the longer the oil will be flowing across Russian territory to Novorossiisk.

THREATS TO POLITICAL STABILITY

There are two principal threats to political stability in the Caspian region: the concentration of power in the hands of authoritarian leaders and the problems posed by secessionist conflicts.[37] Both of these phenomena provide ample

opportunity for Russian political leverage. Russia can keep secessionist conflicts on the boil by feeding weapons to the opposing parties and then use the
simmering conflicts as an excuse to keep its military forces in the region as
peacekeepers. The region's incumbent leaders are all products of the Soviet
nomenklatura, and are known quantities and relatively reliable interlocutors
from Moscow's point of view. In dealing with secessionist conflicts and external threats (such as Islamic fundamentalism), time and again these leaders have
turned to Moscow for military assistance.

One excuse used to justify authoritarian rule is the presence of secessionist
movements. All of the region's countries except Armenia have large ethnic
minorities, ranging from about 10 percent of the population in Uzbekistan to
nearly 50 percent in Kazakhstan (where ethnic Russians, concentrated in the
northern provinces, make up 38 percent of the total population).

U.S. policy assumes that democratization and economic development will
bolster the independence of these states. Democratization in the long term
probably would weaken Russian influence over the region's leaders, but
democratization is progressing at a snail's pace. And the oil wealth, when and
if it arrives, is likely to consolidate, rather than weaken, the power of the
region's authoritarian rulers, as suggested by the experience of oil-rich countries from Nigeria to the Persian Gulf.[38] There is no guarantee that the revenues from energy exports will "trickle down" to society at large and promote
the development of a pluralistic and civil society.[39]

The second threat, secessionist conflict, impacts pipeline considerations in
a number of ways. The pipeline route to Supsa, for example, runs just a few
miles north of the Armenian-occupied zone in western Azerbaijan. Robert
Kocharian, current president of Armenia and the former president of
Nagorno-Karabakh, was reportedly fond of saying "no peace, no oil." Perhaps more importantly, however, energy riches could impact the balance of
power in the South Caucasus. The Armenians fear that if Azerbaijan starts to
build up its oil wealth it will spend the money on arms and will try to retake
Karabakh by force. Still mindful of the 1915 genocide in Turkey, the Armenians see themselves as fighting for national survival. Their economy is in
ruins due to the 1988 earthquake and land blockade imposed by Azerbaijan
in 1992. Financial transfers from the large Armenian diaspora keep Armenia
afloat, along with trade with Iran and U.S. aid. (Armenia receives $100 mil-
lion annually in U.S. aid, the fourth-largest per capita recipient in the world.)

The Karabakh conflict led Washington to impose economic sanctions on
Azerbaijan. Baku is denied bilateral government aid under Section 907 of the
1992 Freedom Support Act, due to Azerbaijan's alleged persecution of
Armenian inhabitants and subsequent economic embargo against Armenia.
The United States believes that if Caspian oil wealth is shared with Armenia,
Erevan can be persuaded to accept a compromise with Azerbaijan. In Sep-
tember 1997, Armenian president Levon Ter-Petrossian responded to U.S.

pressure by indicating that Armenia was ready to make "serious concessions." This hint caused key figures in the Armenian security establishment to desert Ter-Petrossian, who was forced to resign. In recent years, a powerful pro-Baku lobby has emerged in the United States, fueled by a combination of oil corporation dollars and ex–Cold Warriors who would like to see a band of independent and economically robust states emerge between Russia and Iran.[40] (One of the more colorful characters is independent businessmen Roger Tamraz, who donated $177,000 in return for an invitation to a White House "coffee" at which he plugged the Ceyhan pipeline.)[41] Despite the economic sanctions, the Clinton administration has developed a friendly relationship with President Aliyev. It has urged Congress to revoke Section 907 and has done everything it can to lessen the impact of the sanctions.

Azerbaijan's relations with Iran are strained, which increases Baku's sense of isolation and vulnerability to Russian pressure. Azerbaijan was historically part of Persia until it was ceded to Russia in 1818. Around one-third of Iran's population are ethnic Azeris, and some nationalists in Azerbaijan are calling for the secession of Azerbajani regions from Iran. As a result, Iran has developed friendly ties with Armenia and cross-border trade is brisk between the two countries. Iran also has a growing trade relationship with Azerbaijan. In 1998 Iran was the biggest single customer for Azeri oil, buying $180 million worth.[42] But in November 1998, the Azerbaijani Parliament rejected a proposal that would have routed the main oil export pipeline to the Turkish port of Ceyhan across Iranian territory.

Armenia's closest ally is Russia, with whom it signed a renewed security pact in August 1997, leading to the creation of a common air-defense system in 1999. But Erevan is fearful of possible Russian betrayal in return for a share of the Caspian oil riches, and is aware that Moscow objects to the idea of self-determination for Nagorno-Karabakh (for fear of setting a precedent for Chechnya).

Georgia has secessionist problems of its own. It lost control of the provinces of South Ossetia and Abkhazia in 1992, causing 400,000 refugees to flee into the remainder of Georgia. Bitter fighting continued in Abkhazia for another year, with Russian forces supporting the outnumbered Abkhaz forces. Russian assistance explains how 125,000 Abkhaz could defeat the army of Georgia, a nation of 7 million. Russian peacekeepers are still in place in the two regions, and Russia has forced various concessions from Georgia in return for promises to help bring the conflicts to an end. Georgia reluctantly joined the CIS in 1993 and signed an agreement allowing Russia to keep four military bases on Georgian territory. Tbilisi has received little from Russia in return: even promises to facilitate the return of Georgian refugees to Abkhazia have not been implemented. There may also be trouble brewing in the southern Georgian province of Javakhetiya, which sits astride the proposed Ceyhan pipeline route, and where 55 percent of the population of 230,000 are ethnic Armenians.[43]

It seems clear that the Russian military has been fishing in troubled waters in Georgia, manipulating its peacekeeping role to maintain instability in the region. Many Georgians are exasperated with this state of affairs, and hope that more countries—such as Ukraine—can be persuaded to become involved in peacekeeping operations, possibly in Karabakh and Moldova as well as inside Georgia.

As Sabri Sayari points out in chapter 12, hopes that Turkey would enter the area as a power capable of deflecting Russian influence have not yet been realized, in part because of the ongoing struggle within Turkey between secularist and Islamic forces. Also, Russian relations with Turkey remain good, and during a December 1997 visit then–Prime Minister Viktor Chernomyrdin signed a deal to build a gas pipeline across the floor of the Black Sea ("Blue Stream"), which would boost Russian gas supplies to Turkey from 6 billion to 30 billion cubic meters per year by 2010.[44] This project would be a serious competitor for gas exports from the Caspian region, since there are doubts over Turkey's capacity to absorb such a volume of gas imports. Overall, the geography and conflictual geopolitics of Central Asia and the South Caucasus mean that Russia still plays a pivotal role as the dominant power on the region.

IMPLICATIONS FOR RUSSIAN POLICY

Russia's role in the Caspian region is ambiguous. The legacy of the past weighs heavily on the region. Russia conquered these territories in the last century and subjected them to strong central control during the Soviet years. At the same time, Russia is still an important trading partner and provides a geographical link to the outside world.

Russia itself has an ambivalent attitude toward the energy-rich countries of the Caspian Sea Basin. On one hand, those countries are competitors who can flood the world market with additional supplies and who seem keen to bring in foreign multinationals and the political ties that come with them. On the other hand, Russia very much wants to share in the region's energy development projects, whether in terms of supplying management and technology or in providing export routes. Russia has important ties with other countries adjacent to the region (China, Iran, and Turkey) and cooperation rather than confrontation is likely to be a winning strategy both for Russian national interests and for Russian petroleum companies.

The attitude of the states of the South Caucasus and Central Asia toward Russia is also ambivalent. While they have been dominated by Russia for more than a century, and treated (in the case of the Moslem countries) as cultural inferiors, at the same time Russia has been the conduit for exposure to western ideas. Thus the region's leaders, at least in Central Asia, exposed to Russian culture and a Moscow-centric view of the world through decades of

Soviet rule, do not necessarily see a contradiction between orientation toward Moscow and orientation toward the West. Based on a survey of elites in three main Central Asian countries, one study concludes that: "For various reasons [regional elites] had strong feelings about belonging to what they called 'Eurasia' and they did not perceive their countries as a part of Asia or the Third World."[45]

To this day Russia has military bases and its border guards police the "external" frontier in each of the South Caucasus and Central Asian countries, with the exceptions of Azerbaijan and Uzbekistan. Russian forces have played a crucial role in controversial "peacekeeping" operations in Tajikistan and Georgia, although they long ago withdrew from direct involvement in the Karabakh conflict.

Russia's policy is in transition from the military domination that prevailed in the past to a "post-imperial" future that will hopefully revolve around economic cooperation. The organization that replaced the Soviet Union—the Commonwealth of Independent States—is an unwieldy and ineffective structure that has failed to promote economic cooperation. The countries of the Caspian region have been struggling to build networks of international trade, reaching out to new partners such as the United States, Turkey, and China.

Moscow does not want to see rival powers move into what it considers Russia's sphere of influence. Over the past seven years Russia has at times actively interfered to prevent the region's economic development. That situation is slowly changing, as Russian corporations (such as Gazprom and LUKoil) see the potential benefits of involvement in the international consortia operating in the Caspian Basin. These corporations pressed for early Russian involvement in development projects, seeing that the delaying game played by the Ministry of Foreign Affairs would merely work to the advantage of the western oil corporations. Hence Russia has generally pursued a dual track policy—encouraging Russian companies to get involved in projects while simultaneously dragging their feet over resolution of the Caspian Sea status and trying to block efforts to develop new export routes.

With Evgenii Primakov as foreign minister (after January 1996) and later prime minister (August 1998 to April 1999), Russia pursued an actively pro-Iranian policy. It moved ahead with oil deals, completed the nuclear reactor at Bushehr over U.S. objections, and transferred some sensitive missile technology (although the foreign ministry claims this was without its authorization). While this caused tension in the U.S.-Russia relationship, the main theme of U.S. policy remained to engage rather than contain Russia, on the grounds that Russia has not yet "gone bad" and could still be saved for democracy.

The massive Russian assault on Chechnya in the fall of 1999 increased the pressure from those who believed it was time for the United States to take a tougher approach toward Russia and to stop regarding it as a strategic partner. At the same time the appointment of Vladimir Putin as prime minister in

August 1999 and his nomination as official successor to President Yeltsin on December 31, 1999, raised the stakes for western decision-makers. They now faced an intelligent, pragmatic Russian leader who could well be in power for a decade or more. Putin's rise to power increased the incentives for cooperation as opposed to confrontation.

IMPLICATIONS FOR U.S. POLICY

U.S. policy toward the Caspian region involves two somewhat contradictory policy prescriptions: isolating Iran while integrating the new nations of the former Soviet Union into the global economy.[46] Both of these run counter to the interests of Russia, which has been the dominant power in the region for the past 150 years and which has a good relationship with Iran. Genrikh Trofimenko, a leading Russian analyst of U.S. foreign policy, opines that for the United States "the countries of Central Asia have been second after NATO expansion as a field for anti-Russian activity."[47] Yet at the same time, U.S. policy toward Russia is supposedly one of friendly cooperation.

Washington's chosen policy instrument toward Iran is the imposition of sanctions. The past record of sanctions suggests that they rarely succeed in hurting the economy of the target country, but they have some political utility for U.S. domestic political interests—for the Congress in particular. The historical record suggests that sanctions only work around half the time when the goal is to destabilize governments, and less than one-quarter of the time when Washington hopes to advance such objectives as safeguarding human rights or reversing property expropriations.[48] Multilateral sanctions are more effective than unilateral sanctions, but the United States has been unable to persuade the Europeans or other major players to cooperate in the Iranian sanctions regime. With the exception of denying access to highly advanced military technology, the U.S. sanctions against Iran are entirely unilateral.

U.S. efforts to extend sanctions to third country companies through, for example, the Iran-Libya Sanctions Act (ILSA) have provoked fierce reactions from the European Union, which threatened to protest the U.S. actions before the World Trade Organization. On May 15, 1998, at the G8 summit in Birmingham, President Bill Clinton agreed with the EU that ILSA sanctions would not be imposed on the three companies (Gazprom, Malaysia's Petronas, and France's Total) that had announced plans to invest in Iran's South Pars field. In return the EU pledge to step up cooperation in the fight against international terrorism.

The United States has equivocated on whether the current exports of small quantities of Kazakhstani oil and Turkmen gas via Iran will incur sanctions under ILSA. For example, the Turkmen foreign minister came away from meetings in Washington, D.C., in October 1997 convinced that he had been given a "green light" to build a 200-km gas pipeline into Iran, but this was subsequently denied by U.S. officials. Thus one can be skeptical that the United States will be able to achieve its policy goals in the Caspian region. At the same time, Washington seems locked into its policy stance toward the region, so one should not expect any shift in the U.S. position anytime soon.

Iran aside, another factor casting a shadow over U.S. policy is the uncertainty over how important a contribution Caspian oil will make to the global energy market. Even optimistic projections do not see it providing more than 200 million tons of oil a year (about 5 percent of world exports), while pessimistic projections see only 50 million tons annually coming out of the region.[49]

While there is a fairly high degree of consensus among U.S. elites in developing policy toward the Caspian region, there is far from complete unanimity. American energy corporations feel that the U.S. government could be playing a more active role in turning ambitious plans into reality. U.S. oil companies are generally opposed to the current sanctions regime against Iran, but are split on the issue of Russian involvement in Caspian developments. They are generally skeptical of Russian motives and the feasibility of cooperation with Russian companies, but recognize the geopolitical realities of the region may require such involvement.

Within the administration itself there is a bureaucratic division of labor that can at times cause friction. The State Department holds the line on Iran sanctions, but is also a strong proponent of cooperation with Russia. The Energy Department is focused on doing whatever is necessary to bring Caspian oil to market as speedily as possible in the interests of energy security. The Commerce Department sees its role as helping get business for U.S. companies.

The convergence around a single agenda brings a superficial cohesion to U.S. policy in the region. It is based on the heroic assumption that "all good things go together"—that U.S. policy can secure the attainment of democracy, economic development, national independence, conflict resolution, and good relations between neighbors.

However, what happens if U.S. policy fails? It is quite likely that the political obstacles to building export pipelines will prove insuperable; or that these projects will not be commercially viable if the price of oil drops once again. Washington does not at present have a fall-back position for dealing with these contingencies, which leads one to suggest that U.S. policy toward the Caspian states will be fairly rigid for the next several years.

NOTES

1. Yurii Yeremenko, "Bessmyslennost' eksporta dlia Rossii" [Senselessness of exporting for Russia], *Ekonomicheskaia gazeta*, no. 33 (August 1996); and Oleg Cherkovets, "Bananovaia sudba" [Banana fate], *Sovetskaia Rossiia*, January 6, 1996.

2. *Interfax,* December 23, 1998.

3. Konstantin Baskaev, "Rossiiskii TEK vykhodit na avanstsenu borby s neplatezhami" [Russian energy complex at center of struggle with arrears], *Finansovye izvestiia,* March 13, 1997. Oil alone accounts for 20 percent of federal revenue. See *Itar Tass,* September 14, 1997.

4. *Petroleum Economist,* January 14, 2000.

5. Oleg Davydov, "Period liberalizatsii vneshnei torgovli zavershen" [The period of trade liberalization is completed], *Segodnia,* June 6, 1996.

6. *Kommersant Daily,* October 17, 1997.

7. *Finansovaia Rossia,* October 9, 1997.

8. Anders Aslund, "Reform Versus 'Rent Seeking' in Russia's Economic Transformation," *Transition* (OMRI) 2, no. 2 (January 26, 1996): 12–16.

9. William Odom and Robert Dujarric, *Commonwealth or Empire? Russia, Central Asia and the Transcaucasus* (Indianapolis, Ind.: Hudson Institute, 1995).

10. "Vozroditsia li Soiuz?" [Will the Union be Reborn?], *Nezavisimaia gazeta,* May 23, 1996: abridged translation published in *Transition* 2, no. 15 (July 26, 1996): 32–35.

11. Vadim Dubnov, "Gifts from Aliev," *Itogi,* no. 28 (July 15, 1997): 22–23. Some, such as Andranik Migranyan, even place Uzbekistan in this alliance. "SNG nachalo ili konets istorii?" [CIS: The beginning or the end of the story?], *Nezavisimaia gazeta,* March 26, 1997.

12. Neil Malcolm, et al., eds., *Internal Factors in Russian Foreign Policy* (New York: Oxford University Press, 1996).

13. Foreign Ministry legal official Yakov Ostrovskii noted that many governors had signed agreements with foreign countries that "are not always in full compliance with Russia's legislation and international obligations, its political and economic interests." *Interfax,* June 16, 1995.

14. Aleksandr Bovin, "Rossiia vozvrashaetsia v mirovuiu politiku" [Russia returns to world politics], *Izvestiia,* December 26, 1997.

15. Rem Viakhirev, "Ya tak prosto ne sdamsia" [I will not give up that easily], *Nezavisimaia gazeta,* March 25, 1997.

16. LUKoil raised $420 million by selling bonds from 1995 to 1997 and planned another $780 million through global depositary receipts. *Bloomberg News,* April 8, 1997.

17. Lena Johnson, *Russia in Central Asia: A New Web of Relations* (London: Chatham House, 1998).

18. Yakov Pappe, "Neftiannaia i gazovaia diplomatiia Rossii" [Oil and gas diplomacy of Russia], *Pro et Contra* 2 (Moscow: Carnegie Endowment), Summer 1997.

19. Yurii Fedorov, "Kaspiiskaia politika Rossii: k konsensusy elit" [Russia's Caspian policy: toward an elite consensus], *Pro et Contra* (Moscow: Carnegie Endowment) 2 (Summer 1997); and Yurii Fedorov, "Russia's Policies toward Caspian Region Oil: Neo-Imperial or Pragmatic?" *Perspectives on Central Asia* 1 (September–October 1996).

20. John Mitchell, *The New Geopolitics of Energy* (London: Royal Institute of International Affairs, 1996), 63.

21. Michael Wyzan, "Oil, Gas, No Cure for Economic Woes," *Radio Free Europe/Radio Liberty,* January 5, 1999.

22. John Roberts, *Caspian Pipelines* (Royal Institute for International Affairs: London, 1996); Rajan Menon, "Treacherous Terrain: The Political and Security Dimen-

sions of Energy Development in the Caspian Sea Zone" *NBR Analysis* 9 (The National Bureau of Asian Research), no. 1 (February 1998).

23. Turkmen natural gas exports fell from 26 billion cubic meters (bcm) in 1996 to 6.5 bcm in 1997. See Bruce Pannier, "Energy Riches Still Locked Out of World Markets," in *Annual Survey of East Europe and the Former Soviet Union* (Armonk, N.Y.: ME Sharpe, 1998), 404–407.

24. *RIA News Agency*, December 29, 1999; press conference by Transneft President Semen Vainshtok, December 8, 1998, Kremlin International News Service.

25. *Interfax*, June 5, 1997. For half the distance, the existing line will be rebuilt, and for the remainder a parallel line will be built.

26. *Nezavisimaia gazeta*, April 9, 1997.

27. For a relatively sobering view of Kazakhstan's prospects, see chapter 5, "Kazakhstan: The Long-Term Costs of Short-Term Gains," by Pauline Jones Luong.

28. For details on Azerbaijan's energy and foreign policy goals, see chapter 4, "Azerbaijan: The Politicization of Oil," by David Hoffman. See also Menon, "Treacherous Terrain."

29. *Obshchaia gazeta*, March 9, 1997, and *The U.S.-Azerbaijan Chamber of Commerce Virtual Investment Guide 1999*, <www.usacc.org>.

30. *Financial Times*, December 14, 1999.

31. *Moscow Times*, December 7, 1999.

32. Aleksei Chichkin, "Oil and the Chechen War," *Vek*, February 16–22, 1996, p. 7; and Igor Rotar, "Will Caspian Oil Flow over the 'Northern Variant'?" *Prism* (Jamestown Foundation) 3, no. 12, part 3 (July 25, 1997).

33. Robert Ebel, "Geopolitics and Pipelines," *Analysis of Current Events* (New York: Association for the Study of Nationalities, February 1997); and Shabonti Dadwa, "Caspian imbroglio," *Strategic Analysis* 22, no. 5 (1998): 751–60 (New Delhi: Institute for Defense Studies).

34. Jan Kalicki, "U.S. Perspectives on NIS Energy Transportation," October 6, 1999, USIA transcript.

35. Alternative pipeline routes are discussed elsewhere in this book, so I will focus here on Russia's perspective, as it relates to U.S. policies.

36. "Statement of Robert W. Gee, Assistant Secretary for Policy and International Affairs Before the Subcommittee on International Economic Policy, Export and Trade Promotion, Committee on Foreign Relations, United States Senate, on Implementation of U.S. Eurasian Transport Corridor Policy," February 25, 1998.

37. For an overview of the region's politics, see Karen Dawisha and Bruce Parrott, eds., *Conflict, Cleavage and Change in Central Asia and the Caucasus* (Cambridge: Cambridge University Press, 1998).

38. Currently only Kyrgyzstan and Armenia, which do not contain commercially significant quantities of oil and gas, have leaders who came to power through an electoral process.

39. Terry Lynn Karl, *The Paradox of Plenty: Oil Booms and Petro-states* (Berkeley: University of California Press, 1997). See also her "Crude Calculations: OPEC Lessons for Caspian Leaders," chapter 3 in this volume.

40. "Caspian Caviar," *Washington Times*, July 28, 1997; Ariel Cohen, "US Policy in the Caucasus and Central Asia," *Heritage Foundation Backgrounder*, July 24, 1997; and Thomas Goltz, "Catch 907 in the Caucasus," *National Interest*, no. 48 (Summer 1997).

41. *International Herald Tribune*, March 27, 1997.

42. *Radio Free Europe/Radio Liberty*, December 21, 1998.

43. Ugur Akinci, "Javakhetia: the Bottleneck of Baku-Ceyhan Pipeline," *Silk Road* 1, no. 2 (December 1997).

44. Currently Turkey buys Russian gas via a small pipeline across Bulgaria.

45. Rafis Abazov, *Central Asian Republics' Search for a "Model of Development,"* Sapporo: Slavic Research Center Occasional Papers, no. 61 (1998).

46. "U.S. Interests in the Caucasus Region," Hearing Before Committee on International Relations of the House of Representatives (Washington D.C.: Government Printing Office, July 30, 1996).

47. Genrikh Trofimenko, "Tsentral'noaziatskii region: Politika SShA" [Central Asia: U.S. policy], *S.Sh.A.* (IMEMO, Moscow), no. 11 (1998): 21–36. See also Sergei Samuilov, "Politika SshA v otnoshenii SNG" [U.S. policy toward the CIS], *S.Sh.A.* (IMEMO, Moscow), no. 10 (1998): 41–51.

48. Gary Hufbauer, Jeffrey J. Schott, and Kimberly Ann Elliott, *Economic Sanctions Reconsidered,* 2nd edition (Washington D.C.: Institute for International Economics, 1990), 93.

49. Laurent Ruseckas, "Energy and Politics in Central Asia and the Caucasus," *Access Asia Review* (The National Bureau of Asian Research) 2, no. 1 (July 1998).

10

The Afghan Civil War: Implications for Central Asian Stability

Shireen T. Hunter

On February 15, 1989, after more than a decade of war against the Afghan resistance movement, the last Soviet troops withdrew from Afghanistan. Between its military retreat in February 1989 and its disintegration in December 1991, the Soviet Union remained politically engaged in Afghanistan and supported the government of Mohammad Najibullah in Kabul. Nevertheless, the Soviet Union came to accept the idea of a more broad-based and representative government for Afghanistan, provided that it included the communists.[1]

However, several factors made it impossible to set up a broad-based and representative government, including: (1) the Najibullah government's serious lack of legitimacy; (2) diverging interests of various *mujahideen* (Islamic resistance) groups and their regional and international backers; and (3) Pakistan's determination to have a controlling influence in Afghanistan.

The 1979 Soviet invasion of Afghanistan intensified linkages between Central Asia and Afghanistan and had a significant impact on the evolution of political culture in the Soviet Republics. In particular, the conflict helped spread Islamist ideas in Central Asia, especially in Tajikistan and Uzbekistan. The involvement of Central Asian soldiers and officers during the early days of the Afghan war contributed to their national and religious reawakening. Because of these factors, the Soviets soon stopped sending Central Asian personnel to Afghanistan.[2] Moreover, some observers have claimed that U.S., British, Pakistani, and Saudi intelligence services set about recruiting Central Asians for military training at Islamic indoctrination centers in Pakistan's northwest frontier.

The USSR's demise fundamentally altered the geostrategic and political context of governance in Afghanistan. Soon after the Soviet collapse, Russia's leaders withdrew support from Najibullah and he fell from power. In April 1992, the forces of Ahmad Shah Massoud, the ethnic Tajik military commander, entered Kabul. Massoud and his *mujahideen* belonged to the moderate Islamist Jamiat-e-Islami political group led by Burhanuddin Rabbani. Najibullah's efforts to leave the country failed, but Massoud spared his life and he lived under UN protection until Taliban forces hanged him in 1996.

The collapse of the Najibullah government did not result in the reunification of Afghanistan or in the establishment of a broad-based government. Rather, it ushered in a debilitating civil war during which the Afghans have inflicted further human and material damage on their country. For example, rocket attacks launched by the forces of Gulbuddin Hekmatyar of the radical Hezb-e-Islami destroyed most of Kabul and killed 25,000 people. Hekmatyar's motive was to weaken the government of Rabbani, which had assumed power on the basis of the Pakistani-brokered Peshawar agreement.[3]

Post-Soviet international and regional systemic changes further linked Afghanistan's fate with the political and economic future of the Central Asian republics. In the new world order, geopolitical questions related to the exploitation and export of Central Asia's energy resources and the U.S. policy of containing Iran became increasingly important to Afghanistan.

Since 1994 and the end of the post-Soviet honeymoon between Russia and the West, competition for the control of Eurasia has further complicated Afghanistan's political dynamics.[4] The result has been that the Central Asian states have come to see the Afghan conflict as either potentially or actually impacting their own security and other interests. This perception has led some Central Asian countries, notably Uzbekistan, to take a more active role in the Afghan power game.[5] The dynamics of Tajikistan's civil war, which began in 1992, became interconnected with the Afghan conflict, as some Islamist Tajik forces received assistance from Afghan groups and many Tajik refugees fled to Afghanistan.

In 1994, the Taliban emerged on the Afghan scene—fundamentally altering the country's political dynamics. During a period of four years, the Taliban managed to capture and control nearly 90 percent of Afghan territory. Two events particularly stand out in the Taliban's march to victory: their capture of Kabul on September 1996 and of the northern city Mazar-e-Sharif in August 1998 (their first victory in the north in 1997 had proved short-lived).

The rise of the Taliban—an extremist group adhering to a puritanical interpretation of Islam whose members were trained in religious schools in Pakistan and who were recruited from among Afghan refugees and Pakistani nationals—was linked to the following major factors: (1) Pakistan's growing frustration at the inability of its main ally, Gulbuddin Hekmatyar, to unseat the Rabbani-Massoud government, and hence its determination to find an alternative; (2) the hardening U.S. position toward Iran in the context of Washington's "Dual Containment" strategy; (3) the Russo-Western competition for the control of Eurasia, as well as the intensification of competition between Iran and Pakistan in Afghanistan and in Central Asia; and (4) increasing attention paid by the energy companies and the United States to Pakistan as an outlet and market for Central Asia's energy. Since Pakistan has no common borders with Central Asia, Afghanistan as a transit point acquired influence for these actors.

The first step regarding a pipeline through Afghanistan to Pakistan was a 1995 agreement between the governments of Turkmenistan and Pakistan and the Argentinean energy company Bridas to begin studying the possibility of building a pipeline to export Turkmen natural gas via Afghanistan.[6] Shortly afterward, however, the Turkmen government reneged and decided to deal with the American energy concern, Unocal.[7]

The change in the position of the Turkmen government coincided with acceleration of the Taliban's development and its military advances, culminating in the capture of Kabul in 1996. This development and the replacement of Bridas by Unocal as the main company to develop and export Turkmen gas to Pakistan also created the perception among many observers in the region and abroad that the United States supported the Taliban—or at least that it had no objection if its allies, Pakistan, Saudi Arabia, and the United Arab Emirates—did so.[8] As one report alleges, there was more direct involvement by Unocal in the emergence of the Taliban phenomenon.[9]

In short, the rise of the Taliban has been inextricably linked with the politics of Central Asian energy and pipelines. Likewise, pipeline considerations became indirectly related to the Taliban's rise. Because the emergence of the Taliban prolonged the Afghan civil war, it could be argued that the issue of a pipeline through Afghanistan helped exacerbate the conflict. However, because the Afghanistan-Pakistan route for exporting Central Asian energy was never considered a major option for getting these resources—aside from Turkmen gas—to the international marketplace, the impact of the Afghan conflict on the direction of pipeline routes has been less important.

At the same time, the Afghan conflict, and in particular the emergence of the Taliban, has had negative consequences for the Central Asian countries' security environment and other interests, albeit to varying degrees. Tajikistan has been most negatively affected by Afghan events, but since the emergence of the Taliban and the group's extremist brand of Islam and alleged support for Islamists in Central Asia, the security concerns of other Central Asian states have also increased.

Because the Afghan civil war continues, a definitive assessment of the impact that the long conflict has had on the question of pipelines and the security of Central Asian countries is not possible. It is possible, though, to assess the impact that the Afghan conflict has had thus far on the pipeline debate and security in Central Asia; and to examine the likely impact of various scenarios for Afghanistan's evolution on Central Asian security and the energy issues. Before addressing these questions, however, it is important to provide a brief history of Afghanistan's situation through late 1999.

FACTORS BEHIND AFGHANISTAN'S PREDICAMENT

Afghanistan's current problems result from deep-rooted forces related to its social, economic, and political development, its specific ethno-cultural characteristics, and regional and international forces—notably the dynamics of the Cold War and post-Soviet international systems.

Indigenous Factors: Ethno-Linguistic Diversity and the Politics of Modernization

Afghanistan is a multi-ethnic and multi-linguistic country. The two largest groups are the Persian-speaking Tajiks and the Pashtun, who predominantly speak another Eastern Iranian language, Pashtu. The exact percentage of Persian and Pashtu speakers in Afghanistan's total population has never been clear, but estimates of about 45 percent for each group appear realistic (with other linguistic groups making up the other 10 percent).[10]

Most Afghans are Sunni Muslims, but there is a substantial Shi'a minority. In fact, the Shi'a constitute between 15 and 20 percent of the total population, although generally there has been a tendency to underestimate this number.[11] Historically, Afghanistan's Shi'as have been repressed. For example, all public display of Shi'a mourning ceremonies during the month of Muharram—marking the martyrdom of the third Shi'a Imam, Hussein Ibn Ali—have been prohibited. These ethnic and sectarian differences, plus tribal loyalties, have diluted the Afghan sense of national unity. In fact, Afghanistan's name does not denote either a common ethnic root or a cultural notion of identity. The Pashtun have their own ethnic and cultural sense of identity, code of behavior, and values (*Pashtun Wali*), and the Tajiks and other Persian-speakers consider Afghanistan to be the real cradle of ancient Iranian civilization—a belief shared by the pre-Taliban Pashtun elite.[12]

Effects of Modernization

The process of modernization in Afghanistan from the 1920s through the 1970s, including the granting of greater educational and professional opportunities to women and the abolition of the veil, led to new divisions and tensions within Afghan society and elicited reactions from traditionalists. Although the government was successful in repressing such opposition, in the 1960s and 1970s a successful Islamic youth movement developed in Afghanistan. In 1975, this group organized an uprising against the central government in the Panchshir Valley. The revolt was easily put down by the army, and some of the Islamist students, such as Gulbuddin Hekmatyar, were forced to flee to Pakistan, where they were given refuge by the government.[13] Another aspect of Afghanistan's modernization was the penetration of socialist ideas among the Afghan intellectual elite, including in the ranks of the military. This situation resulted from the close relationship that had developed between Afghanistan and the Soviet Union, including between their military establishments, since the 1930s. The openly anti-Islamic and atheist views of the communists who assumed power in 1978, along with their more aggressive secularizing and centralizing policies, further disrupted traditional Afghan social patterns and made Islam the principal alternative to communism and the main medium of resistance to the communists.

However, Islamic resistance groups differed widely in their political and ideological views, goals, and even vocabulary. Organizationally, too, the movement was divided in two ways: first, between the political organizations with headquarters outside of Afghanistan (mostly in Pakistan) and the military groups operating within the country; and second, along ethnic, sectarian, and regional lines.[14]

Regional and International Dynamics: Impact on Afghanistan

Since the late nineteenth century, when it became a semi-independent state, Afghanistan's significance in international relations has been as a buffer against potential intruders into strategically more important regions, such as India, Persia, and the Persian Gulf. Olivier Roy, the well-known French scholar of Afghanistan, is correct in his assessment that the British made Afghanistan a semi-independent state in order to bar Russia's intrusion into India and Persia.[15] Afghanistan's status as a buffer state was maintained after the establishment of the Bolshevik regime and throughout the Cold War—until the Soviet invasion in 1979.

Afghanistan's relations with Pakistan have historically been tense. After India's partition, the British gave Pakistan (instead of Afghanistan) a large segment of Pashtun-inhabited land that formed part of British India. Since then, the Afghans have nurtured irredentist sentiments toward what they called Pashtunistan. Although there were no such territorial disputes between Iran and Afghanistan, disagreements over the distribution of the waters of the Helmand (Hirmand) River and a degree of cultural rivalry have caused tensions between those two countries as well.[16] Afghanistan's relations with Iran and Pakistan were further strained because of Pakistan's involvement in the U.S.-led regional security system (the Baghdad Pact and later the Central Treaty Organization, or CENTO), which was created in the 1950s as part of a series of regional alliances—including the North Atlantic Treaty Organization (NATO) and the Southeast Asian Treaty Organization (SEATO)—to contain the Soviet Union. By the early 1970s, Iran-Afghanistan relations improved when Iran offered Afghanistan financial assistance and easy access to its Persian Gulf ports and signed an agreement on the Helmand waters that was highly favorable to Afghanistan.[17]

During this period, Iran also tried to wean Afghanistan away from Russia and to help resolve the Pakistan-Afghan dispute. The fall of the monarchy in Afghanistan, following a military coup d'état in 1973 led by the king's cousin, Lt. General Muhammad Daoud Khan with the help of pro-Soviet officers, did not alter either Iran's strategy or Afghanistan's desire to expand the range of its partners and policy options. Rather, under Daoud, Afghanistan distanced itself from the Soviet Union and moved closer to Iran and Saudi Arabia, despite the help that the coup leaders had received from the pro-Soviet officers.

By the early 1970s, the Arab presence, especially that of the Saudis and Libyans, had increased in Afghanistan and Pakistan, respectively. Throughout the 1970s, a complicated pattern of cooperation and competition also emerged between Saudi Arabia and Iran: the two countries cooperated in supporting the Daoud regime and weaning Afghanistan away from the Russians, but each was also keen to enhance its own position and to undermine that of the other. Saudi Arabia, especially, was anxious to prevent the Shah of Iran's plan to create a non-Arabic coalition of states in the region.[18] Meanwhile, Pakistan was supporting Afghan Islamists opposed to General Daoud, believing that they would not have irredentist sentiments toward Pakistan's Pashtun-inhabited areas. In short, by the early 1970s, Afghanistan, in addition to being the subject of Cold War rivalry, had become an arena of regional competition.

IRAN, PAKISTAN, SAUDI ARABIA, AND THE AFGHAN WAR:
1980–1991

The Afghan War began at a time when the West had lost Iran. This enhanced Pakistan's strategic position and value for the United States as a barrier to the consolidation of the Soviet presence in Afghanistan, a principal promoter of the Afghan resistance movement, and the main conduit for military and economic aid to the Afghan *mujahideen.* Pakistan's new role cemented its relations with both the United States and Saudi Arabia and provided it with a degree of influence over Afghanistan that it had not previously held. Meanwhile, because of the pressures of the Iran-Iraq War and the existence of pro-Soviet sympathies among certain elements of the Iranian leadership, Iran did not provide adequate support to the Afghan resistance, or even to the Shi'a groups in Afghanistan, and thus alienated many Afghans.[19]

After 1989, three factors convinced Iran that it would be better able to influence Afghan events if the Najibullah government remained a player in Afghan politics: (1) a sharp improvement in Soviet-Iranian relations, illustrated by the visit of Soviet Foreign Minister Eduard Shevardnadze to Tehran in February 1989 and his meeting with the Ayatollah Khomeini; (2) the determination of Saudi Arabia and Pakistan to block a meaningful role for Shi'a groups in arrangements for Afghanistan's future; and (3) the erosion of the ideology-driven aspects of Iran's foreign policy. Thus, Iran indicated that after the withdrawal of Soviet troops the period of Jihad in Afghanistan had ended.[20] Nevertheless, during most of the 1990s, Iran's policy toward Afghanistan remained largely passive and reactive. It was not until the Taliban captured Kabul that Tehran began to support the pro-Iranian groups by providing either money or weapons. Even after this event, Iran showed an extraordinary level of naiveté, believing that Pakistan's influence in Afghanistan would preserve the rights of the Shi'as and the security interests of Iran.

For a period during 1990 and 1991, when the Soviet Union's *glasnost* allowed the expression of long-repressed nationalist feelings in Central Asia—and when, therefore, there was a rise in both pan-Turkist and pan-Iranist sentiments—Iran, Tajikistan, and some Afghan elements tried to establish closer relations among Persian-speaking peoples who shared a common Iranian cultural heritage.[21] However, the ambivalence of Iran's Islamist leaders regarding Iranian nationalism meant that it did not pursue this policy with vigor. Moreover, some of the Islamist forces in Tajikistan had ambivalent sentiments about Iran and notions of pan-Iranism because of their nostalgia for pre-Islamic Iranian traditions. American opposition to Iran also dampened the desire for cooperation with Iran for Persian-speaking, non-Shi'a groups in Afghanistan. Indeed, until the rise of the Taliban, Ahmad Shah Massoud, the Tajik commander, adopted an anti-Iranian posture in order to curry favor with America and to gain its support for the Tajik-dominated government in Kabul.[22]

The Afghan war also helped Saudi Arabia to expand its influence in Afghanistan and Pakistan. It achieved this goal by funding the Afghan *mujahideen* and extensive efforts to spread the Wahhabi school of Sunni Islam. This school is the strictest of Sunni groups and is especially anti-Shi'a. It also does not look favorably on mystical approaches to Islam such as those practiced by various Sufi brotherhoods. During the 1980s, an increasing number of Arab volunteers, financed by Saudi money, poured into Afghanistan and Pakistan, and the Saudis founded Wahhabi schools, which provided full board and a monthly allowance for students.[23]

In addition to creating a new generation of religious teachers and adepts, the Wahhabis waged a campaign against Afghanistan's indigenous Islam—mostly of the Hanafi school—and its Sufi traditions. The Wahhabis offered money to those who destroyed Sufi sacred sites (*zyarats*) and who joined the most extremist of Afghan resistance groups. They even set up local Wahhabi republics in Nuristan, Rech Valley in Kunar, and Argo Valley in Badakhstan.[24] The Taliban are the final and most formidable product of this long-term strategy.

U.S. POLICY: 1979–1991

The United States has played a key role in Afghan developments since the Soviet invasion and will continue to do so in the future. Between the introduction of Soviet troops into Afghanistan in November 1979 and their withdrawal in February 1989, the main goals of U.S. policy were to prevent the consolidation of Soviet power in Afghanistan, to prevent Afghanistan from becoming a staging post for Soviet advances into Iran and the Persian Gulf, and to force the Soviets to retreat.

Until the Soviet military withdrawal, U.S. policy was not much concerned with the post-war Afghan government. After the Soviet Union's agreement to withdraw from Afghanistan, coupled with a sharp improvement in U.S.-Soviet relations, differences emerged within the U.S. government regarding Afghan policy. For Congress, the main concern was to secure continued funding for the *mujahideen,* whereas in the executive branch, the primary issue was whether the United States should cooperate with the Soviet Union and the Soviet-supported Afghan government in order to stabilize the country. According to some observers, the U.S. intelligence community was skeptical about the Soviet Union's intentions, including its sincerity in trying to withdraw from Afghanistan.[25]

Meanwhile, Pakistan and Saudi Arabia were unwilling to deal with either the communist government or those *mujahideen* and political groups that were not their clients. In short, Saudi Arabia and Pakistan wanted to redeem their past investments in Afghanistan by putting in place a government that would be responsive to their interests and would act as a barrier against Iranian influence. Either because of lack of interest or other motives, the United States went along with this approach because it also wanted to isolate Iran.

THE AFGHAN CONFLICT, 1991–PRESENT: FRAGMENTATION, RADICALIZATION, AND INTEGRATION

As has been the case throughout its modern history, since 1991 Afghanistan's destiny has been determined by the interaction between its own indigenous forces and the influence of regional and international dynamics. As would be expected, the consequences of the resistance to the Soviet invasion and the civil war have had considerable social and political outcomes, including:

1. The political fragmentation of society that resulted from the development of various resistance groups and political parties linked with them, accompanied by a weakening of national cohesion and a strengthening of ethnocentric and sectarian identities;

2. Religious radicalization, especially among the Sunnis. This is best exemplified by the rise of the Taliban, but also by some Shi'a groups;

3. Integration along ethnic, linguistic, and sectarian lines. Today, all Persian speakers, whether ethnically Tajik or otherwise, identify more as a separate group, as do the Shi'as, despite their political divisions. The Pashtu-nationalist and religiously radical Taliban, with their anti-Tajik and anti-Shi'a excesses, have contributed to this process. These changes mean that, even if the Taliban were to impose its will by force, the country would remain unstable until some form of decentralized government recognizing these cleavages emerged.

THE COLLAPSE OF THE SOVIET UNION AND THE INTENSIFICATION OF REGIONAL RIVALRIES: IMPACT ON AFGHANISTAN

The Soviet Union's disintegration, combined with the effects of the victory of the U.S.-led coalition against Saddam Hussein in the 1991 Persian Gulf War, drastically altered the character of the international political system and hence the dynamics of regional sub-systems. Among the most significant consequences of the Soviet collapse for the situation in Afghanistan were: (1) a dramatic enhancement in the power of the West, especially of the United States, and its capacity to project this power to far-away places; (2) the strengthening of the position of America's regional allies, coupled with the weakening of anti-U.S. states; (3) the increased U.S. ability to pursue punitive policies toward countries with which it had a conflict of interest because of the elimination of the Soviet counterweight, as reflected in the hardening of U.S. policy toward Iran since 1992; (4) U.S. efforts to prevent the emergence of regional powers or groups of powers which could potentially challenge its supremacy; and (5) the erosion of the cooperative aspects of relations among neighbors of the former Soviet Union and the accentuation of their conflictual dimensions. The worsening of Pakistani-Iranian and Turkish-Iranian relations, along with the intensification of their competition in Central Asia, has been the result of this phenomenon. These systemic developments strongly impacted on the Afghan scene. Most importantly, they led to the rise of the Taliban.

THE ORIGINS AND EVOLUTION OF THE TALIBAN: INTERNAL AND EXTERNAL CAUSES

The emergence of the Taliban has its roots in the Afghan refugee camps of Peshawar, Pakistan, and the Qandhar district of Afghanistan, as well as in the history of the steady spread of puritanical versions of Islam, including

Wahhabism, to Pakistan and Afghanistan. The Taliban has also been an out-come of the post-Soviet geopolitical and geo-economic setting, including the new balance of regional power in the Middle East and South Asia and the strategic and economic priorities of the United States and Pakistan.[26]

The Taliban and their supporters in Pakistan and elsewhere tend to portray a rather romantic vision of their beginnings and their rise to power. Reduced to its essentials, this vision portrays the Taliban, especially their leader Mul-lah Mohammad Omar, as simple-minded and pious religious students and former *mujahideen* fighters who, having become disappointed and disgusted with the inability of the various *mujahideen* groups to pacify the country, set out to stabilize Afghanistan. Furthermore, this version goes, the Afghans, exhausted by continuous strife, welcomed the Taliban. Therefore, the Tal-iban's victories have been the result of this popular fatigue and "widespread unhappiness with infighting, corruption, moral decay and brutality among the existing *mujahideen* parties."[27]

The Taliban's supporters also dismiss their harsh behavior toward women and the strict administration of Islamic justice as being necessitated by Afghanistan's exceptional conditions, and they maintain that there is no doubt that the Taliban will revise their views once they are free from fighting battles.[28]

The reality of the Taliban, however, is that intellectually the movement is the product of decades of Wahhabi proselytizing, which contributed to the strengthening and expansion of the Wahhabi-inspired Deobandi school,[29] whose political arm in Pakistan is the Jamat-e-Ulema-i-Islam. Politically, they are the product of the new, post-Soviet conditions of South and Central Asia, of the post-1992 U.S. strategy to maintain a high level of influence in the energy belt stretching from the Caspian Sea to the Persian Gulf while pre-venting others from dominating the region, and of Pakistan's ambition to dominate Afghanistan and to become a major link between Central Asia and the outside world, including in the field of energy.

Pakistan, the Taliban, and the Pipeline Connection

The decision by Pakistan to train and arm the Taliban and help them gain power was motivated by a variety of factors and aspirations. Pakistan's fore-most objective in creating the Taliban was to put in power in Kabul a gov-ernment that would be responsive to Islamabad and to eliminate Iran's influ-ence. This point is illustrated in a statement by General Nasseerullah Babar, Pakistan's Minister of the Interior, who in late 1993 assumed responsibility for Pakistan's Afghan policy. Babar stated: "I'll see to it that Iran is neutral-ized in Afghanistan."[30] Part of the general's strategy to achieve this goal was the creation of the Taliban.

Long-standing security considerations, especially concern over Afghan irredentism toward Pakistan's northwest frontier, have been the driving force behind Islamabad's policy to put its own man in charge in Kabul. However, Pakistan's desire to become a transit corridor for Central Asia in trade, including a pipeline route, and a major political player in the region should not be underestimated.

Before the issue of a pipeline through Afghanistan became topical, Islamabad tried to entice the Central Asian countries to establish road links through Pakistan.[31] When the U.S. policy of preventing any pipeline from passing through Iran became firmly established, Pakistan became determined that a southern outlet should pass through its territory. But Pakistan suffered from one major handicap, namely the lack of a common border with Central Asia; hence, there was a need to stabilize Afghanistan under its own control. Since Hekmatyar had not been able to achieve this objective, and the Rabbani-Massoud government was not likely to become a client of Pakistan, Islamabad had to create a new force.

Convergence of U.S. and Pakistan Interests

It is doubtful that Pakistan would have been able to achieve its goal of developing and putting in power the Taliban if a degree of convergence with U.S. interests had not emerged. By 1994, a difficult period in U.S.-Pakistan relations (from 1990 to 1993) had come to an end.[32] During this period, some Pakistani military and political leaders, most notably General Mirza Aslam Beg, chief of the Pakistani army and a *bête noire* of the United States, toyed with the idea of a Southwest Asian Security Compact, comprising Iran, Pakistan, and Afghanistan.[33] But these ideas were abandoned, and Pakistan decided to support the U.S. strategy of containing and encircling Iran. The Taliban were to be one important instrument of achieving this goal. This may be why General Beg has claimed that the United States supported the Taliban in order both to undermine Iran's influence in Afghanistan and to drive a wedge between Iran and Pakistan.[34]

The involvement of Unocal in schemes to build pipelines to bring Turkmen gas and later Kazakhstani oil to Pakistan and beyond created another common interest between Pakistan and the United States. The question of whether the United States actively supported the Taliban or, as one ex-U.S. official put it, merely "winked" as Pakistan and Saudi Arabia did the actual work, is controversial and difficult to resolve. However, there have been statements by a number of U.S. officials that indicate at least tacit support for the Taliban. For example, in 1995 U.S. Assistant Secretary of State for South Asian Affairs Robin Raphel told a congressional hearing that Taliban leaders supported a peaceful political process, in contrast to factional leaders who did not want to give up power for the good of all the country. In 1996, after the Taliban's capture of Kabul, Raphel said that the Taliban had to be acknowledged as an "indigenous movement" that had demonstrated "staying power."[35]

Such statements have led some observers to comment that Raphel at times gave the impression of acting as "cheerleader" for the Taliban. Moreover, after the Taliban's entry into Kabul, U.S. officials spoke of seeking early talks with them, and the acting State Department spokesman, Glyn Davies, said that the United States could see "nothing objectionable" about the version of Islamic law that the Taliban had imposed in the areas under their control.[36]

Whether the United States actually supported the Taliban, the posture displayed by U.S. officials of "general ignorance of the identity and origin of the group obfuscating any suggestions that the Taliban were financed and backed by Pakistan" does not seem convincing.[37] Indeed, in view of Afghanistan's role in achieving many U.S. strategic and economic interests, plus the close connection between Pakistan and the United States including their military and intelligence services, it is hard to imagine that the Taliban could have gone as far as they have without U.S. acquiescence.[38]

However, two developments led to a change at least in the tone of U.S. policy toward the Taliban. These were the harsh behavior of the Taliban, especially toward women, which mobilized women's advocacy groups in the United States against the Taliban; and the bombing of U.S. embassies in Nairobi and Tanzania on August 7, 1998, in which Osama bin Laden, an exiled Saudi, a major financial benefactor of the Taliban, and a resident of Afghanistan, was implicated. However, it appeared then, and has become more evident since that time, that the major U.S. concern is to dislodge bin Laden from Afghanistan and that, if the Taliban were to agree to this step, rapprochement would be possible. Indeed, U.S. Secretary of State Madeleine Albright indicated that the Taliban would enhance their chances of American recognition and acceptance by the United Nations if they handed over bin Laden.[39]

The United States also reiterated support for a broad-based government in Afghanistan and the so-called six-plus-two process of stabilizing Afghanistan under UN auspices.[40] But there has been no evidence that Washington worked especially hard to make the process succeed, nor did the United States pressure Pakistan to stop helping the Taliban. Indeed, as of mid-1999, the Taliban, strengthened by large numbers of so-called Pakistani Islamic militants and Arab volunteers, had begun a major offensive to root out the opposition—without any effort on the part of the United States to dissuade the Taliban from doing so. There have even been reports that the United States has begun talks with the Taliban and may have given Pakistan a "tacit nod to bolster its support for the Taliban."[41] Therefore, the inevitable conclusion is that the United States prefers the Taliban to other alternatives, provided that the issue of bin Laden is settled. Indeed, as the Taliban offensive against the opposition began, there were reports that bin Laden would be leaving Afghanistan to find sanctuary elsewhere.[42]

THE AFGHAN CIVIL WAR AND ENERGY DEVELOPMENT: ASSESSING THEIR MUTUAL IMPACT

There is no doubt that there has been a connection between the Afghan Civil War and the issue of pipelines for the export of Central Asian energy. It is clear that the desire to find a southern outlet for Central Asia's oil and gas that would bypass Iran has complicated the Afghan civil war, first and foremost as a factor in the rise of the Taliban. In order to convince the energy companies and the U.S. government that it is a viable outlet, Pakistan had to demonstrate that it could pacify and control Afghanistan and thus guarantee the safety of a prospective pipeline. It is very unlikely that the Rabbani-Massoud government, or any other broad-based government in Afghanistan, would have been sub-servient enough to Pakistan to allow such blanket guarantees to oil companies. Indeed, the Rabbani government would have tried to leverage Afghanistan's position as a transit route to free the country from Pakistani tutelage.

It is important to note here that while the main focus of attention has been on the pipeline to bring Turkmen gas to Pakistan, there were more ambi-tious—albeit not immediate—plans for using Pakistan as a southern transit corridor for other markets. For example, Unocal had plans to extend the gas pipeline to India. At the time, some observers argued that there was already an understanding between India and Pakistan for the export of Turkmen gas to India. Whether such a preliminary agreement existed, Indian and Pakistani nuclear tests in May 1998, tensions between the two countries over Kashmir in the summer of 1999, and the rise of anti-Indian sentiment in Pakistan exclude such options for the foreseeable future.

However, the export of Turkmen gas to Pakistan and eventually to India fits well with the U.S. strategy of weakening Iran, because it preempts the export of Iranian gas to Pakistan and has the added advantage of undermin-ing close Iranian-Turkmen relations, which have not been to Washington's liking.[43] It is useful to remember that at the time the issue of the Afghan pipeline arose, BHP, an Australian energy concern, was working on a feasi-bility study to export Iranian gas to Pakistan and even perhaps to India. According to Ejaz Haider, a Pakistani journalist and political analyst, the United States "scared away" BHP by imposing sanctions on Iran. Moreover, Unocal had plans to construct an oil pipeline from the Turkmen town of Chardzhou, on the border with Uzbekistan, to the coast of Pakistan, through which oil from Western Siberia, Kazakhstan, and Uzbekistan could be exported to world markets. However, the company admitted that before this could happen, large investments would have to be made to develop Pakistan's and Afghanistan's infrastructures, thus making this option very expensive and less competitive than other routes.

It should be noted that Pakistan is a less-than-ideal southern outlet for a number of reasons, including: 1) tense relations with India, which makes the

Indian market less attractive for Central Asian energy; 2) the limited port facility on Pakistan's shoreline; 3) the lack of facilities for the specific purpose of exporting oil and gas; 4) the difficult terrain of Afghanistan; and 5) the instability of Sind and Baluchistan provinces, where Pakistan's two main ports, Karachi and Gwader, are located. In the last two years, two other factors have affected the viability of a southern route, namely: the decision to build a trans-Caspian pipeline west to Turkey; and lower-than-expected growth in global energy demand. Unocal's decision in November 1998 to abandon plans to build a pipeline from Turkmenistan reflect the impact of the above factors, plus the increasingly ambiguous U.S. attitude toward the Taliban in 1998 and 1999. In the long run, if political conditions were to change in Iran to the point of making it an acceptable route, then the Afghan-Pakistan alternative would become even less attractive. By contrast, if Pakistan were to manage to pacify Afghanistan, if the United States were to recognize a Taliban-led government, and if U.S.-Iranian relations remain tense, the idea of a southern route might be revived. However, the developments in the latter part of 1999, notably the continued U.S.-Taliban controversy over bin Laden, the worsening Indo-Pakistani relations, and increased tensions in U.S.-Pakistani ties following a military takeover in Pakistan in October 1999 make the revival of the Pakistan options unlikely.

THE AFGHAN CONFLICT: IMPACT ON CENTRAL ASIAN SECURITY AND THREAT PERCEPTION

The Afghan conflict, both during the period of Russian occupation and the civil war, has had an adverse effect on the security and cohesion of the newly independent Central Asian states. Perhaps the most negative impact of the Afghan war has been the spread of radical Islamist ideas in Central Asia. During the 1980s, the West viewed the brand of Islam practiced in Afghanistan as a strong weapon against the Soviet Union and used it to undermine the hold of communism in the Muslim areas of the Soviet Union. According to noted Pakistani analyst Ahmed Rashid, western intelligence agencies failed to consider the future repercussions of bringing together thousands of Islamic radicals from all over the world, including from Central Asian republics of the Soviet Union.[44]

By 1989, these radicals had established footholds in Tajikistan and Uzbekistan, and they have in the last few years been expanding into Kyrgyzstan and Kazakhstan. In Tajikistan, they contributed to the civil war that bedeviled the country from 1992 to 1997. Even though the opposition and the government reached an agreement under UN auspices in June 1997, Tajikistan's condition is fragile. The Taliban, with their anti-Tajik sentiments, pose a serious threat to Tajikistan. Moreover, since the Taliban forced Massoud out of Kabul, the

Tajikistani government has provided him with a base in the Kuliab region. Therefore, should the Taliban extend their control over northern Afghanistan, Tajikistan may become their next target. At the very least, the Taliban may entice the Muslim opposition to break its agreement with the government, which would reignite the civil war. The Taliban may also want to annex parts of Tajikistan.[45] Even if this were not to happen with a Taliban victory, Tajikistan would be surrounded by two hostile neighbors: Afghanistan and Uzbekistan. This situation would also make it more difficult for Tajikistan to maintain close ties with Iran. The result would be that Tajikistan would become more dependent on Russia, which increasingly is becoming a remote and ineffective protector. Also, pressured by the Taliban's Pashtu nationalism and Islamic puritanism on the one hand, and by Uzbekistan's Turkic nationalism on the other, Tajikistan would find it more difficult to maintain its separate cultural identity. In fact, these two factors may even make the survival of Tajikistan as a separate nation problematic.

Paradoxically, however, in the short term the threat from the Taliban has brought Tajikistan and Uzbekistan closer together. Even Uzbekistan's attitude toward Ahmad Shah Massoud has softened somewhat, and relations between Iran and Uzbekistan have improved.[46] However, it is not clear whether President Islam Karimov of Uzbekistan will be able to overcome his anti-Tajik sentiment and help Massoud more effectively. But even if he were to change his position, it is not certain that this would be enough to prevent a Taliban victory. Uzbekistan felt threatened even by Massoud's version of Islam, which is more moderate than that of the Taliban and even than that of Hekmatyar. Thus with Russian help, Uzbekistan had tried to maintain a kind of secular corridor in the provinces under the control of the ethnic Uzbek General Abdul Rashid Dostum.

After the fall of Kabul in 1996 and especially after the victory of the Taliban in Mazar-e-Sharif in 1998, Uzbekistan joined Russia and Tajikistan in efforts to stop the Taliban's march to victory. Russia first tried to create an anti-Taliban coalition in the context of the collective security program of the Commonwealth of Independent States (CIS). Because of divisions within the CIS regarding the whole concept of collective security, this effort has not been successful. All the same, a more limited anti-Taliban cooperation has developed among Tajikistan, Russia, and Uzbekistan.[47] Kyrgyzstan, although concerned about the speed of radical Islam into its own territory, especially the Uzbek-inhabited areas of Osh and Jalalabad, and despite the infiltration of Uzbek Islamists into its Batken area in August 1999, has adopted a low-key approach toward the Taliban phenomenon. The same has been true of Ka- -zakhstan, which is geographically more remote and to date less vulnerable to the Islamist threat. However, because of the recent events in Kyrgyzstan, Kazakhstan has been taking the Islamist threat more seriously.

Meanwhile, Turkmenistan has embraced the Taliban and has even tried to mediate between it and the Afghan opposition. It has also tried to forge economic and trade links with the Taliban. In March 1999, Reuters reported that Turkmen Foreign Minister Boris Shikhmuradov and Taliban officials agreed to boost trade and other ties. According to the Taliban-controlled radio, Voice of Shariat, Shikhmuradov met the Taliban leader, Mullah Mohammad Omar, in Kandahar. Turkmenistan's approach results from its realization that, so far, its close relations with Iran have not met its energy export and transport needs because of the U.S. determination to exclude Iran from Eurasian pipeline and transport networks. Turkmenistan's decision to opt for the trans-Caspian pipeline, coupled with the reopening of the Pakistan option, illustrates this realization. Moreover, because of the lack of a strong Islamist movement, Turkmenistan feels less threatened by a potential infectious impact of the Taliban.

CONCLUSION AND OUTLOOK

The Afghan conflict, both at the stage of confrontation with the Soviet Union and during the civil war, has had significant and generally adverse consequences for the political stability and security of Central Asian countries, with Tajikistan having suffered most. The international interest in the energy resources of Central Asia and the debate over how to get them to international markets have had a negative impact on the Afghan conflict and have helped to prolong it. However, the impact of the conflict on decisions regarding energy transport routes has been relatively small because of other limitations and difficulties of the Afghan-Pakistan route and the U.S. desire to favor Turkey as the main export hub for Central Asian energy.

The impact that Afghan events will have on future Central Asian security issues and pipeline routes will depend on: (1) how the Afghan conflict is resolved; (2) the fate of the trans-Caspian pipeline; (3) the outlook for energy demand and price; and 4) the future of U.S.-Iran relations. If Afghanistan were stabilized through the establishment of a broad-based government that espoused a more moderate version of Islam, and in which Afghanistan's diverse ethnic and sectarian groups were represented, the security concerns of Central Asian states would be reduced. This development would also revive the idea of at least a limited southern pipeline route through Afghanistan.

If the Taliban manage to subdue other forces and to gain control of the whole region without softening their ideology, the security concerns of the Central Asian states would intensify. In this scenario, even the risk of conflict between a Taliban government and some Central Asian states, particularly Tajikistan, cannot be ruled out, especially if the Taliban actively try to export their ideology to Central Asia and to destabilize the existing governments. However, it is also possible that once in control of all of Afghanistan, the Taliban might try to reach a *modus vivendi* with their Central Asian neighbors and Russia, a step which would help it to pacify its northern front and concentrate on the frontier with Iran.

The degree of influence that Pakistan might be able to exert over a Taliban-ruled Afghanistan would also have a tremendous impact on the future of Afghan-Central Asian relations. If Pakistan could control the Taliban, it would try to moderate the group's behavior and to establish a working relationship between the Taliban and the Central Asian states. Indeed, since spring 1999, Pakistan has been trying to improve its relations with Russia, in part to neutralize Russian opposition to Taliban control of Afghanistan. It was for this purpose that former Prime Minister Nawaz Sharif visited Moscow in March 1999.

Under this scenario, too, the option of a southern route might be resurrected. However, in mid-1999, the Afghan situation still remained fluid, despite some gains by the Taliban, thus making accurate predictions concerning Afghanistan's future, the impact of Afghan events on Central Asia's security, and the future viability of a southern pipeline difficult.

NOTES

1. For a full and excellent discussion of these issues and U.S.-Soviet talks in this respect, see Barnett R. Rubin, *The Search for Peace in Afghanistan* (New Haven: Yale University Press, 1995).

2. On the impact of the Soviet occupation of Afghanistan on Central Asia, see Alexandre Bennigsen and Marie Broxup, *The Islamic Threat and the Soviet State* (London: Croom Helm, 1983).

3. See Amin Saikal, "The Rabbani Government, 1992–1996," in *Fundamentalism Reborn? Afghanistan and the Taliban*, ed. William Maley (New York: New York University Press, 1998), 33.

4. The election of the ultra-nationalist Vladimir Zhirinovsky to the Russian Parliament in October 1993, the rise of anti-western sentiments in Moscow, and Russia's insistence that the post-Soviet space, especially its southern rim, is a special sphere of Russia's influence added to western fears about emerging Russian neo-imperialism. See William E. Odom, "A New Russian Imperialism May Be Coming," *International Herald Tribune*, October 26, 1993. Also see Gillian Tet and Steve Levine, "Russian Peacekeepers Raise Imperialism Fears," *Financial Times*, December 20, 1993.

5. Uzbekistan played this game through Afghanistan's Uzbek minority and, in particular, General Abdul Rashid Dostum. Some observers have noted that Uzbekistan may have wanted to absorb the Uzbek-inhabited parts of Afghanistan. See Anthony Hyman, "Russia, Central Asia, and the Taliban," in *Fundamentalism Reborn?* ed. Maley, 105.

6. "Turkmenistan: Argentine Pipeline Suit Rejected," *Foreign Broadcasting Information Service (FBIS) Daily Report*, FBIS-SOV-98-303, October 30, 1998.

7. Bridas subsequently filed a lawsuit against Unocal in Texas that was dismissed

on the grounds that since Bridas' agreement was with the Turkmen and Afghan (Rabbani) governments, it fell under Turkmen and Afghanistani law rather than that of the United States, and therefore the dispute should be taken up with Turkmen and Afghanistani authorities. See "Turkmenistan: Argentine Pipeline Suit Rejected," *FBIS Daily Report.*

8. Barnett R. Rubin, "Afghanistan: The Taliban and Regional Realignment," *International Affairs* 73, no. 2 (April 1997): 284. In Pakistan, General Mizra Aslam Beg believed that the United States supported the Taliban both to undermine Iran's influence in Afghanistan and to drive a wedge between Iran and Pakistan. Mizra Aslam Beg, "Impacting Forces of Two Revolutions," *Defense Journal,* October 1998.

9. According to Ejaz Haider, a Pakistani journalist and political analyst, Unocal hired ex-U.S. officials as advisers and launched a campaign to win the support of U.S. policymakers for the Afghan route and for the Taliban. See Ejaz Haider, "Pakistan's Afghan Policy and its Fallout," *Central Asia Monitor,* no. 5 (1998): 3.

10. The Tajiks and other Persian speakers who are mostly concentrated in Herat and are called *Farsiwan* or Persians and the Qizilbash constitute close to 35 percent of the population. Other major ethnic groups are the Hazaras, a people of Mongol origin who have been linguistically Persianized, although their dialect has certain peculiarities that set them apart from other Persian speakers. The Hazaras constitute between 8 and 10 percent of the population. Other ethnic groups include the Uzbeks and a variety of smaller Turkic and autochthonous groups, such as the inhabitants of Nuristan (the historic Kafirstan).

11. The Hazaras are all Shi'as and Persian-speaking, but they are ethnically Turkic. Qizilbash are also Shi'a, and there are Shi'a pockets in Herat and in the north. Most Afghan Shi'as adhere to the Twelver sect, but there are also a number of Ismailis or Sevener Shi'as.

12. According to Pashtu folk legends, the Afghans consider themselves as Ben-i-Israel and claim descent from King Saul (whom they call Talut) through a son called Jeremiah, who is said to have had a son named Afghana—hence the name Afghanistan. In other legends, the Pashtuns claim descent from the Copts, who are considered to be related to the Pharaohs of Egypt. However, most historians consider Pashtuns as Indo-European people. This is also the view promoted by the elite of modern Afghanistan. See the 1955 edition of the *Encyclopedia Britannica,* p. 289.

13. On Afghanistan's process of modernization, see Barnett R. Rubin, *The Fragmentation of Afghanistan* (New Haven: Yale University Press, 1995); and Nazif Shahrani, "State Building and Social Fragmentation in Afghanistan," in *The State, Religion, and Ethnic Politics: Afghanistan, Iran, and Pakistan,* ed. Ali Banuazizi and Myron Weiner (Syracuse: Syracuse University Press, 1986), 23–75.

14. On the composition and peculiarities of Afghan resistance movements, see Olivier Roy, *Islam and Resistance in Afghanistan* (Cambridge: Cambridge University Press, 1990).

15. Roy, *Islam and Resistance in Afghanistan.*

16. One cause of the Iran-Afghan tension was the cultural policy of the Afghan government. At one and the same time, the government promoted the Pashtu language and folk culture, highly tinged with anti-Persian overtones, and the establishment of a several-millennia Aryan root for Afghanistan with claims to the effect that the Afghans were the real creators of the ancient and post-Islamic Iranian civilizations.

Needless to say, Iran resented this claim. Meanwhile, with some justification, many Afghans accused the Iranians of suffering from a feeling of cultural superiority.

17. As a junior foreign service officer, I was part of the Iranian delegation, headed by the late Prime Minister Amir Abbas Hoveida, that traveled to Kabul and signed the agreement.

18. The Shah intended to do this by bringing Afghanistan and possibly India into the Regional Cooperation for Development Organization that was formed by Iran, Turkey, and Pakistan in July 1964.

19. Anwar-ul-Haq Ahady, "Saudi Arabia, Iran, and the Conflict in Afghanistan," in *Fundamentalism Reborn?* ed. Maley, 120

20. Barnett R. Rubin, "Afghanistan Under the Taliban," *Current History* 98, no. 625 (February 1999): 86.

21. For a discussion of the tripartite Iran-Tajik and Afghan cooperation and the whole issue of the post-Soviet revival of Pan-Iranism, see Shireen T. Hunter, *Central Asia Since Independence* (Washington, D.C.: CSIS Praeger, 1996), 31–33.

22. Apparently, the United States was not forthcoming toward Massoud because it felt that the Pashtuns would not accept a basically Tajik-led government. On Massoud's efforts to befriend the United States and the American snub, see Anwar-ul-Haq Ahady, "Saudi Arabia, Iran, and the Conflict in Afghanistan," 126.

23. Roy, *Islam and Resistance in Afghanistan*, 218.

24. Roy, *Islam and Resistance in Afghanistan*, 218.

25. Robert Gates, deputy national security advisor and previously deputy to CIA Director William Casey, "remained ultra-skeptical of Gorbachev's intentions," Rubin, *The Search for Peace in Afghanistan*, 99.

26. Wahhabi influence had existed on the Indian sub-continent, and thus in what is now Pakistan, since the nineteenth century. However, traditionally, Wahhabism was not a principal influence on Indian Islam. Growing Pakistani-Saudi relations in the 1960s and 1970s helped spread Wahhabi influence, but it was in the 1980s that it expanded beyond anything experienced earlier, both in Afghanistan and in Pakistan. The Deobandi school of Islam in India was influenced by Wahhabism. Adherents of this school are influential within certain Pakistani Islamic circles, such as the Jamiat-ul-Ulema-i-Islam, which, in turn, has close links to the Taliban.

27. Zalmay Khalilzad, "Anarchy in Afghanistan," *Journal of International Affairs* 51, no. 1 (Summer 1997): 45.

28. Nabi Misdaq, "What Do the Taliban Want in Afghanistan?" *Asian Survey* 1, no. 2 (December 1996): 13.

29. The Deobandi School, which was established in the nineteenth century, developed from conservative reform movements among Indian Moslems. Deobandis looked for inspiration to Shah Walliullah, an eighteenth-century thinker influenced by his contemporary, Muhammad ibn Addul Wahhab of Arabia. Barnett R. Rubin, "Afghanistan: Persistent Crisis Challenges the UN System," *WRITENET Country Papers*, <http://www.unhcr.ch/refworld/country/writenet/wriafg03.htm> (January 29, 2000).

30. Quoted in Haidar, "Pakistan's Afghan Policy and its Fallout," 2.

31. For example, Pakistan had reached agreement in principle with Kyrgyzstan to extend the Bishkek-Tarugar-Kashgar highway and the Karakoram highway to Islamabad and Karachi, and with Uzbekistan to build a highway linking Tashkent to Karachi.

32. For an analysis of U.S.-Pakistan relations in this period see Maleeha Lodhi, "The Pakistan-U.S. Relationship," *Defense Journal,* April 1998.

33. General Mirza Aslam Beg was a principal proponent of this concept, but the roots of the idea go to earlier times. General Muhammad Zia ul-Haq at one time supported such a compact, but in a more extended form that would have included Turkey and the Gulf Arab states. For a discussion of these issues, see Shireen T. Hunter, "South-West Asian Security Compact: Problems and Prospects," in *Dilemmas of National Security and Cooperation in India and Pakistan,* ed. Hafeez Malik (New York: St. Martin's Press, 1993), 34–53; "A Military Consensus with Iran-Afghanistan," *Pakistan Profile,* no. 6 (February 24, 1989); and "The Great Game: Third Time Lucky," *The Economist,* December 10–16, 1988.

34. Richard Mackenzie, "The United States and the Taliban," in *Fundamentalism Reborn?* ed. Maley, 90–103.

35. Mackenzie, "The United States and the Taliban," 91.

36. Mackenzie, "The United States and the Taliban," 95. The author claims that U.S. officials met privately with the Taliban.

37. Mackenzie, "The United States and the Taliban," 90–103.

38. Steve Levine, "Helping Hand," *Newsweek,* October 13, 1997.

39. On the statement of Secretary Albright, see "The Taliban's Strategy for Recognition," *The Economist,* February 6, 1999; and on broader U.S. policy toward the Taliban, see the testimony of Karl F. Inderfurth, U.S. Assistant Secretary of State for South Asian Affairs, before the Senate Subcommittee on Foreign Operations on March 9, 1999. Inderfurth stated that the United States supported the six-plus-two process under UN auspices.

40. The six-plus-two group includes the states bordering Afghanistan—China, Iran, Pakistan, Tajikistan, Turkmenistan, Uzbekistan—plus Russia and the United States.

41. "The Taliban Launch A Summer Offensive," *Stratfor Commentary,* <http://www.stratfor.com>, July 28, 1999.

42. Amir Zia, "Afghanistan Fighting," *Reuters,* July 28, 1999; and "Bin Laden Reportedly Leaving Afghanistan," *Washington Post,* July 30, 1999.

43. For more on relations between Turkmenistan and Iran, see chapter 6, "Turkmenistan's Energy: A Source of Wealth or Instability?" by Nancy Lubin.

44. Rasid, Amed, "Islam in Central Asia" in Rould Sagdeen (ed.), *Islam in Central Asia,* forthcoming.

45. On the Taliban's territorial designs and desire to help the Tajik opposition see "Tajikistan: Afghanistan's Taliban Said To Help Tajik Opposition," *FBIS Daily Report,* FBIS-SOV-96-195, October 4, 1996.

46. Ahmad Shah Massoud visited Uzbekistan in the context of the UN-sponsored Afghan peace talks—the six-plus-two process—held in Tashkent in July 1999. In view of the past mistrust between him and Uzbek leaders, this is an important development. "Afghan General Ahmad Shah Mas'ud visiting Uzbekistan," *FBIS Daily Report,* FBIS/SOV-1999-0721, July 21, 1999; and "Iran's Foreign Minister Arrives in Tashkent," *FBIS Daily Report,* FBIS-NES-1999-0505, May 5, 1999.

47. See: "Russia: Yeltsin—Russia, Tajikistan, Uzbekistan can hold Taleban" *FBIS Daily Report,* FBIS-SOV-98-226, August 13, 1998.

11

China's Interest in Central Asia: Energy and Ethnic Security

Dru C. Gladney

China is often portrayed as expanding beyond its borders in search of economic and energy security, with Central Asia considered to be one of its most immediate areas of conquest. Figuring large in these scenarios are China's restive border minorities, particularly the Muslim people known as the Uighurs, and the role they may play in either obstructing China's expansionism or in derailing China's economic development drive. For China, three separatist issues continue to plague its foreign and domestic policy: Taiwan, Tibet, and Xinjiang. Indeed, China's strong objections to the NATO involvement in Kosovo were specifically voiced as a protest against aiding and abetting separatists. This highlights the concerns that China has toward outside support for its internal separatists groups. Since 1993, China's own domestic energy supplies have become insufficient for supporting modernization, increasing its reliance upon foreign trading partners to enhance its economic and energy security, leading toward the need to build what Chinese officials have described as a "strategic oil-supply security system" through increased bilateral trade agreements. By examining recent relations between the newly independent Central Asian states and China—especially as they relate to Xinjiang Province—this chapter will shed light on some of the future security concerns for China as they relate to energy supply and other regional issues.

RUPTURES AND RELATIONS

In 1997, bombs exploded on two buses on March 7 (killing two), and in a city park in Beijing on May 13 (killing one); as well as in the northwestern border city of Urumqi, the capital of the Xinjiang Uighur Autonomous Region on February 25 (killing 9). The following year, more than thirty other bombings occurred, six in Tibet alone. Most of these are thought to have been related to demands by Muslim and Tibetan separatists. Eight members of the Uighur Muslim minority were executed on May 29, 1997, for alleged bombings in northwest China, with hundreds arrested on suspicion of taking part in ethnic riots and engaging in separatist activities. Though sporadically reported since the early 1980s, these incidents have been increasing since 1997 and documented in a recent scathing report of Chinese government policy in the region by Amnesty International.[1] A report in the *Wall Street Journal* of the arrest in August 1999 of Rebiya Kadir, a popular Uighur business woman during a visit by the United States Congressional Research Service delegation to the region, indicates China's random arrests have not diminished since the report, nor does China appear to be concerned with western criticism.[2] These campaigns, according to the April 1999 Amnesty International Report, have led to 210 capital sentences and 190 executions of Uighurs since 1997.[3] The proliferation of websites documenting the "plight of the Uighurs" and developments in the region indicates increasing international interest and support. Significantly, despite on-going tensions and frequent reports of isolated terrorist acts, there has been no evidence that any of these actions have been geared to disrupt the economic development of the region. Most confirmed incidents have been directed against Han security forces, recent Han Chinese immigrants to the region, and even Uighur Muslims perceived to be too closely collaborating with the Chinese government. The following discussion of the Uighurs will attempt to suggest why there have been increasing tensions in the area and what the implications are for future international relations.

Most analysts agree that China is not vulnerable to the same ethnic separatism that split the former Soviet Union. But few doubt that should China fall apart, it would divide, like the USSR, along centuries-old ethnic, linguistic, regional, and cultural fault lines.[4] If China did fall apart, Xinjiang would split in a way that, according to Anwar Yusuf, president of the Eastern Turkestan National Freedom Center in Washington, D.C., "would make Kosovo look like a birthday party."[5] These divisions showed themselves at the end of China's last empire, when it was divided for over 20 years by regional warlords with local and ethnic bases in the north and the south, and by Muslim warlords in the west. Hence, for Central Asia, the break-up of the USSR did not lead to the creation of a greater "Turkestan" or a pan-Islamic collection of states, despite the predominantly Turkic and Muslim population of the region. Rather, the break-up fell along ethnic and national lines. China clearly is not about to fall apart, not yet anyway. Yet it also has ethnic security concerns that directly relate to its energy development plans, and this interrelationship is growing in importance.

THE UIGHUR AND PROSPECTS
FOR AN INDEPENDENT UIGHURSTAN

Chinese histories notwithstanding, Uighurs firmly believe that their ancestors were the indigenous people of the Tarim Basin, which many today refer to as Uighurstan, a vast region that did not become known in Chinese as "Xinjiang" ("new dominion") until the eighteenth century. Nevertheless, the identity of the present people known as Uighur is a rather recent phenomenon related to Great Game rivalries, Sino-Soviet geopolitical maneuverings, and Chinese nation-building. While a collection of nomadic steppe peoples known as the "Uighur" have existed since before the eighth century, this identity was lost from the fifteenth to the twentieth centuries. It was not until the fall of the Turkish Khanate (552–744 AD) to a people reported by the Chinese historians as Hui-he or Hui-hu that we find the beginnings of the Uighur Empire. At this time the Uighur were but one collection of nine nomadic tribes, who initially in confederation with other Basmil and Karlukh nomads, defeated the Second Turkish Khanate and then dominated the federation under the leadership of Koli Beile.[6] China's relations with the Uighur were part of their larger contacts with the traditional peoples and places of Central Asia along the ancient Silk Road.

The region today known as Xinjiang received that label in 1759 when the region was finally brought under Qing control, a dynasty established by the Inner Asian Manchus, who were concerned to rid the region of continued Mongolian control. Even Manchu control was short-lived in the region, disrupted by Taiping and Uighur rebellions, Russian influence, and finally its own collapse in 1911. It is clear from the history of China's interactions with Central Asia that the country generally has sought only economic opportunities and secure borders. Attempts at military domination of the region, and indeed the incorporation of Xinjiang, Mongolia, and Tibet, only took place under foreign dynasties in China. Then, as now, there would be great risks to China's internal solidarity were it to attempt territorial expansion into Central Asia.

Conquest of the Uighur capital of Karabalghasun in Mongolia by the nomadic Kyrgyz in 840 led to the dispersion of the early Uighur kingdom. One branch that ended up in what is now Turpan took advantage of the unique socioecology of the glacier fed oases surrounding the Taklamakan desert and were able to preserve their merchant and limited agrarian practices, gradually establishing Khocho (or Gaochang), the great Uighur city-state based in Turpan for four centuries (850–1250). With the fall of the Mongol empire, the decline of the overland trade routes, and the expansion of trade relationships with the Ming Dynasty, Turpan gradually turned toward the Islamic Moghuls and, perhaps in opposition to the growing Chinese empire, adopted Islam by the mid-fifteenth century.

The Islamization of the Uighur from the tenth to as late as the seventeenth centuries gradually displaced their Buddhist religion and
Manichaean practices, but did little to bridge widely dispersed oases-based
loyalties. From that time on, the people of "Uighurstan" centered in Turpan, who resisted Islamic conversion until the seventeenth century, were
the last to be known as Uighur. The others were known only by their oasis
or by the generic term of "Turki." With the arrival of Islam, the ethnonym
"Uighur" fades from the historical record. That Islam became an important, but not exclusive, cultural marker of Uighur identity is not surprising
given the sociopolitical oppositions with which the Uighur were confronted. In terms of religion, the Uighurs are Sunni Muslims, practicing
Islamic traditions similar to their co-religionists in the region. In addition,
many of them are Sufi, adhering to branches of Naqshbandiyya Central
Asian Sufism.

The end of the Qing dynasty and the rise of Great Game rivalries between
Russia and Great Britain saw the region torn by competing loyalties and
marked by two short-lived and drastically different attempts at independence: the short-lived proclamations of an "East Turkestan Republic" in
Kashgar in 1933 and in Yining (Ghulje) in 1944.[7] These rebellions and
attempts at self-rule did little to bridge competing political, religious, and
regional differences within the Turkic Muslim people who became known
officially as the Uighur in 1934 under successive Chinese Kuomintang war-
lord administrations.[8] Enormous ethnic, religious, and political cleavages
from 1911 to 1949 pitted Muslim against Chinese, Muslim against Muslim,
Uighur against Uighur, Hui against Uighur, Uighur against Kazakh, war-
lord against commoner, and nationalist against communist.[9] This extraordinary factionalism caused a huge depletion of lives and resources in the
region that still lives in the minds of the population. Indeed, it is this mem-
ory that many argue keeps the region together, a deep-seated fear of widespread social disorder.

Today, despite continued regional differences along three, and perhaps
four, macro-regions, including the northwestern Zungaria plateau, the southern Tarim Basin, the southwest Pamir region, and the eastern Kumul-Turpan-
Hami corridor, nearly eight million people spread throughout this vast region
regard themselves as Uighur among a total population of sixteen million.[10]
Many of them dream of, and some militate for, an independent Uighurstan.
The "nationality" policy under the Kuomintang identified five peoples of
China, with the Han in the majority. The Uighur were included at that time
under the general rubric of "Hui Muslims" that included all Muslim groups
in China. This policy was continued under the communists, who eventually
recognized 56 nationalities: the Uighur and 8 other Muslim groups split out
from the general category "Hui" (which was confined to mainly Chinese-
speaking Muslims).

The Uighur designation in China, therefore, continues to mask tremendous regional and linguistic diversity and also includes many "non-Uighur" groups such as the Loplyk and Dolans that had very little to do with the oasis-based Turkic Muslims that became known as the Uighur. At the same time, contemporary Uighur separatists look back to the brief periods of independent self-rule under Yakub Beg and the Eastern Turkestan Republics, in addition to the earlier glories of the Uighur kingdoms in Turpan and Karabalghasaun, as evidence of their rightful claims to the region. Contemporary Uighur separatist organizations based in Istanbul, Ankara, Almaty, Munich, Amsterdam, Melbourne, and Washington, D.C., may differ on their political goals and strategies for the region, but they all share a common vision of a unilineal Uighur claim on the region, disrupted by Chinese and Soviet intervention.

The independence of the former Soviet Central Asian Republics in 1991 has done much to encourage these Uighur organizations in their hopes for an independent Uighurstan, despite the fact that the new, mainly Muslim Central Asian governments all signed protocols with China in Shanghai in spring 1996 that they would not harbor or support separatist groups. These protocols were reaffirmed in the August 25, 1999 meeting between then–Russian President Boris Yeltsin and Chinese Premier Jiang Zemin, who committed the "Shanghai Five" nations (China, Russia, Kazakhstan, Kyrgyzstan, and Tajikistan) to respecting border security and the suppression of terrorism, drug smuggling, and separatists.[11] On June 15, 1999, three alleged Uighur separatists (Hammit Muhammed, Ilyan Zurdin, and Khasim Makpur) were deported from Kazakhstan to China, with several others in Kyrgyzstan and Kazakhstan awaiting extradition.[12]

Within the region, though many portray the Uighur as united around separatist or Islamist causes, Uighur continue to be divided from within by religious conflicts (including competing Sufi and non-Sufi factions), territorial loyalties, linguistic discrepancies, commoner-elite alienation, and competing political loyalties. These divided loyalties where evidenced by the attack in May 1996 on the Imam of the Idgah Mosque in Kashgar by other Uighurs, as well as by the assassination of at least six Uighur officials in September 1998. It is also important to note that Islam was only one of several unifying markers for Uighur identity, depending on those with whom they were in cooperation at the time. For example, to the Hui Muslim Chinese in Xinjiang, numbering over 600,000, the Uighur distinguish themselves as the legitimate autochthonous minority, since both share a belief in Sunni Islam. In contrast to the formerly nomadic Muslim peoples, such as the Kazakh, numbering more than 1 million in Xinjiang, Uighurs might stress their attachment to the land and oasis of origin. In opposition to the Han Chinese, the Uighur will generally emphasize their long history in the region. This suggests that Islamic fundamentalist groups such as the Taliban in Afghanistan will have only limited appeal among the Uighur. It is clear that this contested understanding of history continues to influence much of the current debate over separatist and Chinese claims to the region.

Unrest in the Xinjiang Uighur Autonomous Region may lead to a decline in outside oil investment and revenues, while current projects are already operating at a loss. Continued disappointments have diminished expectations over oil exploration in the Tarim Basin, with the entire region yielding only 3.15 million metric tons of crude oil annually, only a small fraction of China's overall output of 156 million tons in 1998. The World Bank loans over $3 billion a year to China, investing over $780.5 million in 15 projects in the Xinjiang region alone, with some of that money allegedly going to the Xinjiang Production and Construction Corps (XPCC), which human rights activist Harry Wu has claimed employs prison (*laogai*) labor. Hearings in the U.S. Senate on World Bank investment in Xinjiang have led Assistant Treasury Secretary David A. Lipton to declare that the Treasury Department would no longer support World Bank projects associated with the XPCC. International companies and organizations from the World Bank to Exxon may not wish to subject their employees and investors to social and political upheavals. China also recently cancelled plans to build an oil pipeline from Kazakhstan to Xinjiang and central China due to the lack of outside investment and questionable market returns.

China's international relations with its bordering nations and internal regions such as Xinjiang and Tibet have become increasingly important not only for economic reasons, but also for China's desire to participate in international organizations such as the World Trade Organization and Asia-Pacific Economic Cooperation (APEC) forum. Though Tibet is no longer of any real strategic or substantial economic value to China, it is politically important to China's current leadership to indicate that they will not submit to foreign pressure and withdraw their iron hand from Tibet. Uighurs have begun to work closely with Tibetans to put political pressure on China in international fora. In an April 1997 interview in Istanbul, Ahmet Türköz, vice-director of the Eastern Turkestan Foundation that works for an independent Uighur homeland, noted that since 1981 meetings had been taking place between the Dalai Lama and Uighur leaders, initiated by the deceased Uighur nationalist Isa Yusup Alptekin.[13] The elected leader of the Unrepresented Nations and People's Organization based in The Hague, an organization originally built upon Tibetan issues, is Erkin Alptekin, the son of the late Isa Alptekin. These international fora cannot force China to change policies, any more than can the annual debate in the U.S. Senate over the renewal of China's Most-Favored Nation status. Nevertheless, they continue to influence China's ability to cooperate internationally. As a result, China has sought to respond rapidly, and often militarily, to domestic ethnic affairs that might have international implications.

CHINA AND CENTRAL ASIAN RELATIONS: CONTEMPORARY CONNECTIONS AND CONTRADICTIONS

Since the breakup of the Soviet Union in 1991, China has become an important competitor for influence in Central Asia and is expected to serve as a counterweight to Russia. Calling for a new interregional "Silk Route," China is already constructing such a link with rails and pipelines. The ethnicization of several Central Asian peoples and their rise to prominence as the leading members of the new Central Asian states will mean that economic development and cross-border ties will be strongly influenced by ancient ethnic relations and geopolitical ties.

In a flurry of trade negotiations and investments in late 1998 and early 1999, the China National Petroleum Corporation (CNPC) pledged more than $8 billion for oil concessions in Sudan, Venezuela, Iraq, and Kazakhstan—plus $12.5 billion to lay four oil and gas pipelines (total length 13,500 kilometres) from Central Asia and Russia to China. Though China recently scaled back plans for these pipelines due to lack of international investment, these actions nevertheless indicate China's willingness to pay more than market value for energy supplies that will enhance its security. According to one report, China offered as much as 30 percent above estimated value for Kazakhstan's Uzen oil fields and more than twice as much as the next highest bidder for its Venezuelan oil fields.[14] China's interests in enhancing its energy security is driving its increasing trade with Central Asia. The growing interdependence of the region is indicated by the fact that trade between Xinjiang and the Central Asian republics has grown rapidly, reaching US$950 million in 1998, and the number of Chinese-Kazakh joint ventures continues to rise, now approaching 200. Xinjiang exports a variety of products to Kazakhstan, as well as to Uzbekistan, Kyrgyzstan, Russia, and Ukraine. Increased economic cooperation with China is providing Central Asia with additional options for markets, trade routes, and technical assistance.

Cross-border ethnic ties and interethnic relations between Xinjiang and Central Asia continue to have tremendous consequences for development in the region, which shares many similarities. Muslims, mostly Uighur, comprise nearly 60 percent of Xinjiang's population. Being Turkic, the Uighurs share a common Islamic, linguistic, and pastoralist heritage with the peoples of the Central Asian states. The Uighurs and other Turkic groups in the region are also closer culturally and linguistically to their Central Asian neighbors than they are to the Han Chinese. The Han (the official majority nationality of China), are also relatively recent immigrants to Xinjiang. The beginning of this century marked an enormous movement of Russian and Han Chinese settlers to outlying Central Asian regions. From 1949 to 1979, China sent Han professionals to Xinjiang to help "open the Northwest." In

1990, estimates put the Han Chinese at 38 percent of Xinjiang's population, up from five percent in 1949, with the Uighur population dwindling to 48 percent. While Russian populations have begun to decline in parts of the Commonwealth of Independent States (CIS) since independence, the Han migration to Xinjiang continues to escalate.

Opportunities in Xinjiang's energy sector attract many of the migrants. China's rapidly growing economy has the country anxiously developing domestic energy sources and looking abroad for new sources. China is expected to import as much as 30 percent of its oil in the year 2000. As China develops into a modern economy, it should see a rise in energy demand comparable to that experienced in Japan, where demand for natural gas and other energy needs has quadrupled in the past 30 years. This is particularly why China has begun to look elsewhere to meet its energy needs and Li Peng signed a contract in September 1997 for exclusive rights to Kazakhstan's second largest oil field. It also indicates declining expectations for China's own energy resources in the Tarim Basin. Once estimated ten years ago to contain 482 billion barrels, today even the president of CNPC admits that there are known reserves of only 1.5 billion barrels.

China hopes to make up for its dependence on Kazakhstan oil by increasing trade. China's two-way trade with Central Asia has increased dramatically since the government in Beijing opened Xinjiang to the neighboring region following the collapse of the Soviet Union in 1991. By the end of 1992, formal trade had jumped by 130 percent; total border trade, including barter, is estimated to have tripled. Ethnic ties have facilitated this trading surge: those with family relations benefit from relaxed visa and travel restrictions. Large numbers of "tourists" from Kazakhstan, Tajikistan, and Kyrgyzstan make frequent shopping trips to Xinjiang and return home to sell their goods at small village markets. Xinjiang has already become dependent on Central Asian business, with the five republics accounting for the majority of its international trade in 1998.

Most China–Central Asia trade is between Xinjiang and Kazakhstan (Xinjiang's largest trading partner by far). From 1990 to 1992, Kazakhstan's imports from China rose from just under 4 percent to 44 percent of its total. About half of China-Kazakh trade is on a barter basis. Through 1995, China was Kazakhstan's fifth largest trade partner, behind Russia, Holland, Germany, and Switzerland. China's trade with Kyrgyzstan also has increased rapidly. Through 1995, Kyrgyzstan was Xinjiang's third largest trading partner, after Kazakhstan and Hong Kong. As early as 1992, China ranked as Uzbekistan's leading non-CIS trading partner. Since then, bilateral trade has increased by as much as 127 percent per year, making Uzbekistan China's second largest Central Asian trading partner. This may be one of the most promising economic relationships developing in Central Asia. The large and relatively affluent Uzbek population will eagerly purchase Chinese goods once remaining border restrictions are relaxed and better transportation links are built. Bilateral trade with Tajikistan increased nearly nine-fold from 1992 to 1995. However, with much of Tajikistan recently in turmoil and the country suffering from a deteriorating standard of living, trade dropped by half in 1996. Trade between China and Turkmenistan has also risen rapidly. China is expected to eventually import Turkmen gas to satisfy the growing energy requirements in the northwest corner of the country. It is clear that while trade figures have fluctuated somewhat between China and the CIS in the last six years, the overall trend is increasing, particularly with Kazakhstan.

While the increasing trade between Central Asia and China is noteworthy, it reflects China's rapidly growing trade with the entire world: trade with Central Asia increased by 25 percent from 1992 to 1994; during the same period total Chinese trade increased almost twice as fast. In fact, during 1995, only .28 percent of China's $280.8 billion overseas trade involved the five Central Asian republics, about the same as with Austria or Denmark. Despite the small trade values, China is clearly a giant in the region and will play a major role in Central Asia's foreign economic relations. For example, China's two-way trade with Kazakhstan is greater than Turkey's trade with all five Central Asian republics. This is so even though predominantly Muslim Central Asia has been a high priority for Turkey.

Multinational corporations are beginning to play a larger role in the development of the region. In Kazakhstan, for instance, foreign firms are estimated to control more than 60 percent of electric power output. A proposed Turkmenistan-China-Japan natural gas pipeline, part of the envisaged "Energy Silk Route" that would connect Central Asia's rich gas fields with Northeast Asian users, demonstrates the potential for cooperation among countries. But it also highlights the growing importance of international companies—in this case Mitsubishi and Exxon—in financing and influencing the course of oil and gas development in the region. With a potential price tag of $22.6 billion, this pipeline—as well as many smaller and less costly ones—would not be possible without foreign participation. Hence, as Michael Mandelbaum argues in chapter 2 of this volume, the new Great Game in Central Asia involves many more players—both governmental and nongovernmental—than the largely two-way Great Game of the nineteenth century. Yet these new international corporate forces do not supersede local ethnic ties and connections that extend back for centuries.

Landlocked Central Asia and Xinjiang lack the road, rail, and pipeline infrastructure needed to increase economic cooperation and foreign investment in the region. Oil and gas pipelines still pass through Russia, and road and rail links to other points are inadequate. A new highway is planned from Kashgar, Xinjiang, to Osh, Kyrgyzstan, to facilitate trade in the area. At the same time, China is planning a new rail link between Urumqi and Kashgar.

Despite increasing investment and many new jobs in Xinjiang, the Uighurs and other ethnic groups complain that they are not benefiting as much as recent Han immigrants to the region. As noted above, this is a major factor in recent Uighur Muslim activism. They insist that the growing number of Han Chinese not only take the jobs and eventually the profits back home with them but that they also dilute the natives' traditional way of life and leave them with little voice in their own affairs.

KAZAKHSTAN-CHINA PIPELINE

The recent "discovery" of oil and gas reserves in the Central Asian Republics has created great interest in the West. Reports of reserves that match or exceed those in the Middle East have recently been tempered by prohibitive exploration and development costs as well as political instability and potential distribution difficulties. High sulfur content of Kazakhstan's oil reserves and the declining expectations over Tarim Basin oil in Xinjiang has meant increasing cost estimates in meeting China's growing energy needs.

The proposed pipelines shipping oil and gas from Central Asia to China for domestic consumption or re-export have several problems that are yet to be resolved. The first is the requirement for foreign investment. Neither China nor Kazakhstan have the funds to build the pipeline. Without substantial loans from the developed nations, the viability of the project is questionable. The second problem is the technical difficulty of constructing a pipeline. The remoteness of Xinjiang coupled with the lack of adequate roads and railways to transport materials make the project even more daunting. Already, Kazakhstan has experimented with shipping small amounts of crude for refinement in Xinjiang by train. This has been expensive and a token attempt at meeting China's energy needs. The third problem China and Kazakhstan face is the separatist movement within Xinjiang that might dissuade foreign investors from putting money into the pipeline. Addressed above, this is a psychologically important influence, though practically speaking, there is little evidence to date that separatists intend to disrupt development projects.

Kazakhstan is interested in building a pipeline through China for several reasons. Kazakhstan needs to find distribution means independent of Russia. The pipeline would provide a means of getting oil and gas to market independent of Russia; create an economic alliance with China that could be used as a counter balance against Russian coercion; and increase its foreign currency reserves through loans and, later, sale of oil or gas. Thus, Kazakhstan is not interested in supporting any group that might jeopardize the construction of the pipeline, and it may use the separatist movement as a proxy to leverage certain concessions from the Chinese. This is a potentially dangerous move as the pipeline would pass through Xinjiang and thus be highly vulnerable if completed. Kazakhstan's best move probably would be to encourage China to defuse the tensions by introducing social reforms that could stabilize the region.

The pipeline from Kazakhstan to China is one of many projects that Beijing is pursuing to ensure continued growth. Thus, for China, the construction of the pipeline does not hold the same significance that it does for Kazakhstan. While China has been a net importer of oil for several years, coal continues to be the primary resource for energy production.[15] The relative independence from gas and oil and the fluctuating global oil prices mean that China is not desperate to build the pipeline, though for environmental considerations China is searching for alternatives to coal.

The pipeline is important for the United States but hardly a vital concern. The production levels of oil and gas in Central Asia have not reached the point to be important to the United States. The United States is interested in the stability and economic development of the region and in ensuring that a mutually beneficial relationship is established with the Central Asian republics. Because the Central Asian region of the CIS shares borders with China, Russia, and Iran, these newly independent states are important to the United States with or without oil. But if oil production were to increase substantially, major distribution lines were to be built, and the U.S. relationship with Russia, China, and the Middle East were to deteriorate further, Central Asia would be of critical strategic importance. Just as importantly for Washington, alternate sources of hydrocarbons for China would mean decreasing reliance on the Middle East as a sole source, thus decreasing competition in the region and the potential for tensions in the Persian Gulf.

In recent years, the United States has welcomed any process that promotes the establishment of free-market democracies. In this regard, a pipeline to China could help to bring Kazakhstan into the global economy, as well as to wean it from sole dependence on Russia.

Even if the United States were interested in helping the separatist movement—as many Uighur separatist organizations hope—it is virtually powerless to do so. Economically, the United States is unable to levy sanctions against China. As one of our top trading partners, an embargo would hurt the United States at least as much as China in the long run. Militarily, the United States lacks the ability to influence Chinese actions in Xinjiang. America could supply the independence movement with small arms, but similar attempts in the past have shown the United States' ineptitude at supporting independence movements because Washington generally lacks the patience, dedication, and political will required. Given recent problems in Kosovo, policymakers are likely to rethink support for any separatist movement, anywhere.

RUSSIA, PIPELINES, AND UIGHUR SEPARATISM

Russia has the most to gain and least to lose by helping the separatists. With the greatest proven reserves in the world, Russia has little to lose if Central Asia cannot extract oil from the region. Despite the Central Asian Battalion (CENTRASBAT) paratroop exercises led by U.S. military forces, Russia knows America's ability to project and support forces in the region is extremely limited. Thus the investment by U.S. oil companies to explore, extract, and develop oil and gas fields is susceptible to Russian influence.

Without a pipeline, the Central Asian states do not have a means of getting energy resources out of the region to western and Asian markets. Thus Kazakhstan and the other nations will remain dependent on the old system of selling oil and gas to Russia for local currency or exchanges of goods while Russia sells to the West for foreign currency. Just as the United States opposes the pipeline from Turkmenistan running through Iran because it will diminish American control, Russia is opposed to pipeline routes that escape Russian influence.

Supporting a Uighur move for independence could conceivably assist Russian policy in a number of ways. First, it would create friction between China and Kazakhstan. Both countries look to profit from the pipeline and have agreed not to encourage secessionist movements. If such a movement were to strengthen, both governments would be forced to reexamine their policies. Second, a serious independence movement would cause China to realign its military assets to take care of the region, which would reduce assets in other Sino-Russian border areas. Additionally, any move that would curtail civil rights in Xinjiang would exacerbate Chinese-American relations, especially if it led to large-scale human rights abuses inflicted on Muslims. It should be noted that most of the Uighur language and research programs in Kazakhstan throughout the 1960s, 1970s, and 1980s were part of a prolonged propaganda campaign that was funded by Moscow to encourage Uighur separatism and attempt to destabilize China. After independence, Kazakhstan strictly curtailed those efforts and greatly reduced the Uighur Research Institute of the Kazakhstan Academy of Sciences.

Deteriorating living standards are increasing tensions among the more than one hundred ethnic groups living in Central Asia. The densely populated Ferghana Valley, home to many of the region's ethnic groups, has been the site of clashes over jobs, land, and natural resources (especially water) since the 1980s. In June 1989, Meskhetian Turks, who had been exiled to the area by Stalin, were attacked by Uzbeks and Tajiks. Another skirmish followed a year later between Uzbeks and Kyrgyz in Osh. There is particular concern about the Tajik-Uzbek conflict, given serious tensions between the two groups and their proximity. One million Tajiks live in Uzbekistan, while both Tajikistan and Kyrgyzstan have sizable Uzbek populations in their parts of the Ferghana Valley. In Kazakhstan, Russian-Kazakh tensions remain high, with 60 percent of Kazakhstani Slavs and Germans still considering their homeland the USSR, not Kazakhstan. Throughout Central Asia, Russians, mostly technicians and other professionals who came after the 1917 revolution, make up roughly one-fifth of the population. Their fears that growing nationalism in countries of the region may become increasingly anti-Russian has prompted many of them to return to their homeland. Efforts to build a Kazakhstani identity have failed to bridge Slavic-Turkic and Orthodox-Muslim differences.

THE NEW GREAT GAME, THE OLD SILK ROUTE, AND THE FUTURE

Central Asia, China, and indeed much of Asia will continue to be shaped by historical forces, policies, and economic development that have brought them closer together in the last few years than in the thirty since the breakdown in Sino-Soviet relations in the early 1960s. Historically, however, we have seen that Central Asia has always been an important cross-roads and meeting place in the heart of Eurasia. It is now reassuming this role in the international marketplace.

The post–Deng Xiaoping leadership of China must seek new solutions to its old ethnic problems in the region. Deng's many crackdowns on separatist movements in the borderlands (he was the political officer for the 1959 PLA army that advanced on Tibet) no longer make sense in a country trying to open itself to world markets and meet global expectations. China must go beyond its former two-pronged policy in the border areas: Political repression coupled with economic investment. Not only has erecting a "steel Great Wall" against separatists in Xinjiang, to use Regional Party Secretary Wang Lequan's terms, failed to keep them out, but it can no longer hide China's problems from the world.[16]

China's are the last Muslims under communism. With the independence of the largely Muslim nations of former Soviet Central Asia, the end of the war in Bosnia, the Israeli-PLO rapprochement, and even the recent peace accords with Muslim separatists in Chechnya and the Philippines, followers of Islam worldwide have begun to pay increasing attention to China's Muslims. China cannot ignore the fact that support for the Bosnian Muslims was the only issue upon which the Iranian, Saudi, and Turkish governments could agree. Turks, through a modern process of globalization and nationalism, see themselves as directly linked to their "brothers" in China and Central Asia. Muslims, through the global "Umma" (the religious belief that all Muslims are one family) and Islamization, also see themselves linked to the region. These international connections and ethno-religious ties will continue to shape and influence China–Central Asian relations. Islam will continue to be a factor that unites and divides many Central Asian states and those they interact with, particularly China, Pakistan, India, and the predominantly Muslim Southeast Asian states of Malaysia and Indonesia.

PRESENT AND FUTURE IMPLICATIONS FOR CHINA–CENTRAL ASIA RELATIONS

This study of the Central Asia–China connection suggests the following conclusions regarding energy and ethnic security:

1. It is clear that while China is seeking to extend its sphere of influence as a regional power in order to enhance its own economic and energy security, it will not seek territorial expansion into the regions of Central Asia. As in the past, the Central Asian states will continue to be viewed as "buffer zones" between the larger Eurasian powers, as well as sources for trade.

2. As such, Central Asian states will honor agreements to insure border security and not harbor separatist activities, yet they may not be able to prevent popular and private support for such activities.

3. Energy will continue to be the economic incentive for cooperation and the most contentious issue between the Central Asian states, Russia, and China, as well as Iran, Turkey, Japan, and the United States. China is willing to strain many of these bilateral relationships in order to enhance its own energy security interests.

4. The indigenous peoples in the region, many of whom live on both sides of the borders between China and Central Asia must realize the benefits of state-sponsored development programs or serious social unrest will continue. Islamic fundamentalist tendencies in the region will increase or fall in direct relation to Muslim participation in state-sponsored development programs.

5. International illegal migration will continue to be a vexing problem for all bordering states, and domestic internal migration will continue to impede development projects, especially for Han Chinese to Xinjiang and Russians out of Central Asia into Russia.

6. Ethnic relations between majorities and minorities in the region will continue to be strained as long as serious inequities and social stratification along ethnic lines persist. Cross-border smuggling of weapons, drugs, and contraband will continue until peaceful resolutions can be achieved in Afghanistan, Kashmir, and Tajikistan, and economic development is more uniformly distributed between ethnic groups in the region.

NOTES

1. Amnesty International, *People's Republic of China: Gross Violations of Human Rights in the Xinjiang Uighur Autonomous Region*, ASA 17/18/99, April 21, 1999.

2. Ian Johnson, "China Arrests Noted Businesswoman in Crackdown in Muslim Region," *Wall Street Journal*, August 18, 1999, p. 1.

3. Amnesty International, *People's Republic of China: Gross Violations of Human Rights in the Xinjiang Uighur Autonomous Region*, 1.

4. Dru C. Gladney, "China's Ethnic Reawakening," *Asia-Pacific Issues*, no. 18 (1995): 1–8.

5. Anwar Yusuf, president of the Eastern Turkestan National Freedom Center, Washington, D.C., interview with author, April 14, 1999. Note that due to this fear of widespread civil disorder, Mr. Yusuf indicated the Eastern Turkestan National Freedom Center did not support a free and independent Xinjiang. Mr. Yusuf met with President Clinton on June 4, 1999, to press for fuller support for the Uighur cause. Turkestan News & Information Network, press release, June 8, 1999.

6. For an excellent historical overview of this period, see Herbert Franke and Denis Twitchett, *Cambridge History of China: Volume 6. Alien Regimes and Border States (907–1368)* (Cambridge: Cambridge University Press, 1994).

7. The best discussion of the politics and importance of Xinjiang during this period by an eyewitness and participant is Owen Lattimore, *Pivot of Asia: Sinkiang and the Inner Asian Frontiers of China and Russia* (Boston: Little, Brown, 1950).

8. Linda Benson, *The Ili Rebellion: The Moslem Challenge to Chinese Authority in Xinjiang, 1944–1949* (Armonk, NY: M.E. Sharpe, 1990).

9. These cleavages are described in considerable detail by Andrew Forbes in *Warlords and Muslims in Chinese Central Asia* (Cambridge: Cambridge University Press, 1986).

10. Justin Jon Rudelson, *Oasis Identities: Uighur Nationalism Along China's Silk Road* (New York: Columbia University Press, 1998), 8. For Uighur ethnogenesis, see also Jack Chen, *The Sinkiang Story* (New York: Macmillan, 1977), 57; and Dru C. Gladney, "The Ethnogenesis of the Uighur," *Central Asian Survey* 9, no. 1 (1990): 1–28.

11. Rym Brahimi, CNN Correspondent, "Russia, China, and Central Asian leaders pledge to fight terrorism, drug smuggling," August 25, 1999 (cited in www.uygur.org /enorg/wunn99/990825e.html).

12. "Kasakistan Government Deport Political Refugees To China" (cited in: www.uygur.org/enorg/reports99/990615.html).

13. Author's interview with Ahmet Turkoz, April 7, 1997, Istanbul.

14. Ahmed Rashid and Trish Saywell, "Beijing Gusher: China pays hugely to bag energy supplies abroad," February 26, 1998, in Turkestan News & Information Network 3, no. 135, June 10, 1999.

15. Between 1991 and 1994, coal supplied 74 percent of China's energy needs. See *China Energy Annual Review, 1994.*

16. *Xinjiang Daily*, December 16, 1996.

12

Turkey's Caspian Interests: Economic and Security Opportunities

Sabri Sayari

The emergence of the newly independent states in the South Caucasus and Central Asia has created new opportunities for Turkey to pursue its economic and political interests. Since the end of the Cold War, Ankara has sought to establish closer ties with the region, particularly with Azerbaijan and the Central Asian Turkic republics. One of the principal objectives of Turkish policy has been to play a major role in Caspian energy development. Faced with a rapidly growing demand for oil and natural gas, Turkey has endeavored to increase its energy supplies through bilateral and multilateral energy agreements with the Caspian states. At the same time, Ankara has spent considerable diplomatic effort to promote the Baku-Ceyhan pipeline project as the principal transit route for the export of Azerbaijani oil to western markets. Geopolitical considerations, economic and political expectations, and environmental concerns have shaped Turkish policy toward the so-called "Caspian Pipeline Derby."

Turkey's strategy to assert its economic and strategic presence in the Caspian area is part of a broader trend in Turkish foreign policy in the post–Cold War period. Since the early 1990s, Ankara has pursued its political and economic interests through an activist foreign policy in the Middle East, the Balkans, and Central Asia.[1] Following the collapse of communism and the end of the perceived Soviet threat, this new activism was prompted by several factors. Turkish officials worried about their country's geostrategic importance for the West. Their concerns were accentuated by Europe's apparent reluctance to admit Turkey as a full member of the European Union. In addition, the rise of political instability, war, and ethnic conflict near Turkey's borders in the Middle East, the Caucasus, and the Balkans prompted Ankara to become involved in these regions to an unprecedented degree in recent history.

In the Middle East, the outbreak of the Gulf War in 1990 and 1991 proved to be a turning point for this new activist approach. Turkey departed from its traditional policy of non-involvement in regional conflicts, took a strong stand against Saddam Hussein's regime, and participated in the U.S.–led Allied Coalition. At the same time, the Gulf War contributed to the emergence of the Kurdish problem as a major, region-wide issue that, in turn, intensified Turkey's own conflict with the ethnic Kurdish rebel organization, the Kurdistan Workers Party (PKK). Turkey's periodic military incursions into Northern Iraq in pursuit of the PKK, and its new strategic relationship with Israel—prompted partly by Ankara's desire to end Syrian support for the PKK—underscored its preference for proactive policies to pursue its political and strategic interests.

In the Balkans, the eruption of ethnic conflicts involving Muslim communities following the disintegration of Yugoslavia had a strong resonance in Turkey due to historic ties dating back to several centuries of Ottoman rule in Southeastern Europe. Ankara responded to the plight of the Bosnian and Albanian Muslims through intense diplomatic efforts on their behalf in western capitals and with international organizations such as NATO. Since the Dayton Peace Accords in 1995, Turkey has participated in the multilateral peacekeeping force in Bosnia-Herzegovina, and a small contingent of Turkish F-16s was deployed in Italy for use in NATO's air campaign against the Milosovic regime during the Kosovo conflict in 1999.

In Central Asia and the South Caucasus, the collapse of the Soviet Union and the emergence of the Turkic republics presented Turkey with another opportunity to expand its regional role. During the Cold War, Turkey's relations with Central Asia and the Caucasus were almost nonexistent. This changed rapidly with the disintegration of the former Soviet Union: in the early 1990s, Turkey embarked on an ambitious effort to establish political, economic, and cultural ties with the region, especially with the new Central Asian Turkic republics and Azerbaijan.[2] By the end of the decade, relations between Turkey and most of the regional states had grown rapidly, encompassing a variety of political, cultural, and commercial networks. Although its earlier expectations to become the dominant regional force proved to be premature and unrealistic, Turkey nevertheless emerged, along with Russia and Iran, as one of the key political actors in the Caucasus and Central Asia.

While this new activist foreign policy has helped Turkey to underscore its regional importance, it also has created new security and political dilemmas for the Turkish government. This particularly has been the case in the Caucasus, where the rise of violent ethnic conflicts and protracted political instability has the potential to spill over into Turkey. Secessionist conflicts in Chechnya and Georgia have ramifications for Turkey's own Kurdish separatist problem. Domestic pressures stemming from the sympathy of the Turkish public for Turkic and/or Muslim groups, concurrent with Ankara's desire to remain neutral and play a moderating role in regional conflicts, have created difficult policy choices.

Moreover, Turkey's increased involvement in the Caucasus and Central Asia has brought into sharp focus regional power rivalries among Russia, Iran, and Turkey.[3] In the unstable regional environment of the post–Cold War period, each of these states sought to expand its political influence and limit the role of others. In particular, there was a discernible rise in tensions between Moscow and Ankara over a number of issues that highlighted their geopolitical concerns regarding each other's policies and behavior. There was apprehension in Moscow about Turkey's efforts to capitalize on its ethnic and cultural affiliations with the Turkic states and possibly alter the geostrategic balance to its favor.[4] Similar perceptions prevailed in Ankara about the reassertion of Russian power and influence in its "near abroad."[5] In particular, the Turkish government worried about: (1) the restationing of Russian troops along its borders with Georgia and Armenia; (2) the flow of military aid from Russia to Armenia; (3) Moscow's efforts to instigate domestic instability in Georgia and Azerbaijan to prevent them from asserting their independence; and (4) Russia's desire to maintain its near-monopoly on the flow of Caspian energy to western markets.[6] Turkish perceptions and security concerns about Russian policies have deep historical roots: the Caucasus region was the scene of major wars and conflicts between the Ottoman Turks and the Russians for more than three centuries. The consolidation of Russian control in the region during the first two decades of the twentieth century was accompanied by a great deal of human tragedy involving various Turkic and Muslim communities.[7] Stalin's territorial demands in Eastern Turkey and the Bosporus Strait at the end of World War II renewed traditional Turkish perceptions about the "threat from the north." Throughout the Cold War period, Turkey sought to protect itself against the perceived Soviet threat by forging close political and military ties with the West, particularly the United States.

TURKEY AND CASPIAN ENERGY DEVELOPMENT

Energy became a major issue in Turkish domestic politics and foreign policy for the first time following the world oil crisis of the early 1970s. Total energy consumption increased threefold between 1962 and 1977, resulting in a sharp rise in Turkey's dependence on oil imports. Following the global oil crisis in 1973, Turkey's annual oil bill tripled in four years to $6 billion, nearly 60 percent of the country's total foreign currency earnings in 1977.[8] Faced with a severe financial crisis, the government decided to reduce energy consumption. This decision led to daily power cuts in major cities that lasted several hours. Industrial production was disrupted and "families shivered in their homes for lack of heating oil, while gargantuan queues of vehicles lined up at the pumps for the meager supplies of petrol."[9] This experience has had a lasting impression on Turkish policymakers; since then, governments of all political persuasions have considered energy to be vital for political stability, economic growth, and national security. They have also viewed energy as an important factor in domestic politics, since the political backlash from blackouts often undermines the popularity of governments and political parties. The oil crisis of the 1970s established a direct link between energy issues and foreign policy for the first time in the republic's history. Until then, Turkey had maintained cordial if not very close relations with most of the Arab countries in the Middle East. In the aftermath of the worldwide oil shock, Turkey intensified its diplomatic efforts to improve relations with the major oil-producing Arab states and adopted a more pro-Palestinian position on the Arab-Israeli conflict.[10]

Although energy remained a major public policy issue during the two decades following the 1973 oil crisis, it has assumed unprecedented importance in the 1990s for two major reasons. First, the country's demand for energy grew rapidly as a result of industrialization, high economic growth rates, and population increases. Oil imports have increased more than 30 percent since 1990, reaching 22 million tons of crude oil annually. This figure is expected to increase to more than 40 million tons by the year 2010.[11] Before the Gulf War, Turkey's main source of crude oil was Iraq. Since the imposition of the UN sanctions on Iraq, Turkey has increased its oil purchases from Saudi Arabia and Iran. At present, Turkey receives nearly two-thirds of its crude oil from these two countries.[12] Turkey's own domestic oil production (which meets about 12.1 percent of domestic demand) has declined in recent years as a result of the natural depletion of fields. Moreover, exploration in the southeast has been affected by the PKK's campaign of political violence and terrorism.

Turkey's need for natural gas has grown at an even more striking pace since it began consuming natural gas in 1976. This trend largely owes to the increased use of gas-fired power plants to satisfy growing industrial and household consumption of electricity. Demand for natural gas is projected to increase significantly over the next twenty years—with rapid industrialization that will require massive amounts of new power capacity. It is estimated that annual gas imports will rise from 8 billion cubic meters (bcm) in 1995 to at least 32 bcm and perhaps as much as 43 bcm by 2010.[13] Since the beginning of gas imports in 1987, Russia has been Turkey's principal supplier of natural gas and currently provides more than 70 percent of total imports.

MANEUVERING OVER PIPELINE ROUTES

The second reason for the growing Turkish focus on Caspian energy development in the post-Cold War environment concerns pipeline construction projects. The transport of Caspian energy to western markets has been a major source of international political and economic maneuvering and diplomacy. Producing states (Azerbaijan, Kazakhstan, and Turkmenistan), neighboring countries (primarily Russia, Turkey, and Iran), western oil companies, and the United States have actively sought to influence the choice concerning the routing of new pipelines. In this intense competition, the Turkish government has actively promoted the Baku-Ceyhan project for the transport of the main Azerbaijani crude oil to world markets. The Turkish proposal envisages the construction of a 1,730-mile pipeline from Azerbaijan across Georgia and Turkey to the Turkish Mediterranean port of Ceyhan at a cost estimated to be approximately $2.7 billion.[14]

Turkey has supported the Baku-Ceyhan route for environmental, geopolitical, and economic reasons. The Turkish government's officially stated case for the Baku-Ceyhan project rests mainly on environmental and safety concerns. Turkey has maintained that if the main export oil from the Caspian is carried through the northern route (from Baku to Novorossiisk and then to the Mediterranean by tanker), the increased shipping and tanker traffic through the narrow Bosporus Strait would create a major environmental problem due to the possibility of collisions and oil spills. There is the fear that increased tanker traffic carrying Caspian crude oil would endanger the lives of Istanbul's 10 million inhabitants, since the 19-mile Bosporus Strait runs through the center of the city.[15] Turkish officials contend that the Baku-Ceyhan route offers an environmentally safer alternative to the northern route, which has been strongly promoted by Russia.

While Turkey's official stand on the Caspian pipeline issue is based mainly on the environmental argument, there is no question that geopolitical concerns have played an important role in Ankara's efforts to secure the construction of the Baku-Ceyhan pipeline. The politics of Caspian pipeline construction have been intertwined with the competition for political and economic influence in the South Caucasus and Central Asia between Russia, Turkey, and Iran. Turkey has viewed the Baku-Ceyhan route as a valuable strategic and political asset that would highlight the country's position as an energy bridge between the emerging Caspian Basin supply centers and western markets. In particular, a major oil pipeline from Azerbaijan through Georgia to Turkey would create closer ties with both of these states that are also within Russia's and (in the case of Azerbaijan) Iran's potential sphere of influence. Many Turks, especially the country's political and military elites, have tended to view this project as a matter of great national and international prestige. The Baku-Ceyhan project has proven to be an effective morale-building mechanism and a source of national pride, thanks in no small part to highly sensationalized press coverage and the efforts of political leaders who have sought to capitalize on this issue for political gain.[16]

Economic considerations also have been important in Turkey's support for a pipeline that would carry Azerbaijani oil across Turkey. Ankara expects to gain economic benefits through transit fees, new jobs, contracts for engineering firms, and a supply of oil at the least cost.[17] Due to the UN sanctions on Iraq, the large terminal facilities at the Mediterranean port of Ceyhan, to which twin pipelines transport oil from northern Iraq, have remained largely idle for nearly a decade. Thus the sanctions on Iraq have also adversely affected the economy of Ceyhan and surrounding areas. Turkish officials believe that the construction of the Baku-Ceyhan pipeline would be a major step toward revitalizing economic life in and around Ceyhan.

Although Washington has endorsed the idea of multiple pipelines, it also has provided Turkey with strong backing for the Baku-Ceyhan project. The United States has become increasingly involved in Caspian energy issues, and its support for Baku-Ceyhan is part of Washington's policy of promoting an "East-West corridor" for Caspian oil and gas pipelines. Washington's preference for Baku-Ceyhan stems partly from a desire to support Turkey—a loyal NATO ally and a strategic country for U.S. national interests—at a time when it faces a strong challenge from Islamist forces at home. U.S. policy is also partly due to Washington's concerns about Russia and Iran. While the Clinton administration has been careful not to antagonize Russia in Caspian energy politics, it has nevertheless sought to prevent Russia from using heavy-handed tactics to assert its power and influence. At the same time, Washington's support for an "east-west energy corridor" underscores its desire to exclude Iran from becoming a potential transit state for the export of Caspian oil and gas.

Until the mid-1990s, Turkey sought to promote Baku-Ceyhan as the principal pipeline for the transport of large volumes of Azerbaijan's oil. Turkish diplomatic initiatives were directed at western governments, especially the United States, and oil companies, and focused on the negative political, economic, and environmental consequences of the northern route sponsored by Russia. Competition rather than cooperation with Russia best characterized Turkish policy. However, in the latter part of the decade Turkey has modified its stand on Caspian pipeline issues. The construction of the pipeline from Baku to Ceyhan still remains a top priority in Turkish foreign and security policy. But it is no longer seen as the only or even necessarily the first possible route for the main oil pipeline. There is growing recognition of Russian interests and of the possibility that there will be multiple pipelines. More importantly, Turkish policymakers have belatedly realized that the commercial concerns and interests of the oil companies are critical for a major pipeline construction project such as Baku-Ceyhan. Western oil companies have been reluctant to make a commitment to Baku-Ceyhan because of its relatively high cost and lack of financing by countries involved in the project.[18] In addition, the uncertainties concerning the actual amount of petroleum reserves in the Caspian Sea and the opening of two early-oil routes, Baku-Novorossiisk and Baku-Supsa, during 1999 have diminished the commercial attractiveness of Baku-Ceyhan for the oil companies. To meet the demands of the Azerbaijan International Operating Company (AIOC) for incentives and protection against cost overruns to make the project commercially attractive, the Turkish government has offered a $2.7-billion cost guarantee for the pipeline and agreed to pay any excess over this price. The Turkish offer also set 2005 as the date for the start of construction of the pipeline from Baku to Ceyhan. It remains to be seen whether this new Turkish policy initiative will in fact result in the completion of the Baku-Ceyhan project during the first decade of the twenty-first century.

EXISTING ENERGY SECTOR AGREEMENTS

Turkey's growing focus on Caspian energy issues in the 1990s led to its participation in the AIOC and a number of major new agreements for the purchase of natural gas. In 1994, the Turkish Petroleum Corporation (TPC) joined AIOC, a consortium of oil companies and the government of Azerbaijan that was formed to extract oil from three of the Caspian Sea fields. TPC, a state-owned company, holds 6.75 percent equity in the AIOC.

To meet its growing demand for natural gas, Turkey signed several new gas purchase agreements with Iran, Russia, and Turkmenistan. In August 1996, Turkey and Iran agreed to a $23 billion natural gas deal whereby Iran would deliver 4 bcm of gas to Turkey annually, with a possible increase to 10 bcm following the completion of a new pipeline connecting the two countries. In December 1997, Russia and Turkey signed an agreement for the "Blue Stream" project, in which the Russian gas company, Gazprom, in a joint venture with Italy's ENI, would construct a 750-mile pipeline under the Black Sea that would provide Turkey with 16 bcm of gas by the year 2010. A year later, in October 1998, Turkish President Suleyman Demirel and his Turkmen counterpart Saparmurat Niyazov signed an agreement for the export of Turkmen gas to Turkey and European markets through a trans-Caspian pipeline that would go under the Caspian Sea, across Azerbaijan and Georgia, and into Turkey. According to this agreement, Turkey has guaranteed to purchase up to 16 bcm of Turkmen gas annually, and another 16 bcm may transit Turkey en route to Europe.[19]

This series of new major gas pipeline and purchase agreements highlights several trends in Turkey's approach to Caspian energy issues. First, Ankara is willing to explore all possibilities for potential new sources of natural gas, since Turkish officials are worried about the likelihood of not meeting the country's growing demand in the near future. Second, Turkey's search for new gas imports is matched in its intensity by competition between the principal suppliers of natural gas (Russia, Turkmenistan, and Iran) for a share of the lucrative Turkish market. Third, Turkish energy policies toward Russia and Iran display both competitive and cooperative tendencies. Moscow and Ankara have competed to promote their own preferred policy choices for the construction of the main oil pipeline from Azerbaijan. This competition has been a source of tension between the two countries. Similarly, Turkey has viewed Iran as a potential rival in Caspian pipeline politics, since an oil pipeline through Iran could be an alternative to Baku-Ceyhan. However, competition and rivalry with Russia and Iran over Caspian oil pipelines has been accompanied by cooperation with the two countries with respect to natural gas. In addition, there are strong commercial relations with both countries, especially between Russia and Turkey. Fourth, driven largely by the urgency to find new gas imports, Turkey has displayed considerable inconsistency in its Caspian energy policies. For example, the Turkish government has often stated its desire to diversify the sources of its gas imports and lessen its dependence on Russia. Yet this concern did not prevent Turkey from signing the Blue Stream agreement, which, if it becomes a reality, could in fact increase Russia's share of the Turkish natural gas market and bolster its status as Turkey's most important supplier. Moreover, the Blue Stream proposal runs counter to the U.S.-backed trans-Caspian pipeline project for the transport of Kazakhstani and Turkmen energy to the Turkish and European markets. By supporting both the Blue Stream and trans-Caspian proposals, Turkey has sent mixed signals concerning its policy objectives. Moreover, given the strong U.S. endorsement of the trans-Caspian route, Turkey's new gas deal with Russia has created tensions between Ankara and Washington at a time when Turkey counts on continued U.S. support for the Baku-Ceyhan project.

THE TURKISH PERSPECTIVE ON REGIONAL INSTABILITY AND GEOPOLITICAL COMPETITION

Political instability and ethnic conflict near its borders in the Caucasus pose a major problem for Turkey with respect to its energy and pipeline interests in the Caspian region. They also vastly complicate Ankara's relations with the three South Caucasus republics—Armenia, Azerbaijan, and Georgia—and with two other important regional actors, Russia and Iran. Since the end of the Cold War, the maintenance of peace and stability in the Caucasus has been a major objective of Turkish foreign policy.[20] Ankara has endeavored to play a stabilizing role in a volatile region through bilateral diplomatic efforts and participation in multilateral attempts at conflict resolution. However, a number of factors, including the lingering effects of historical enmities, contemporary geopolitical rivalries, and domestic pressures at home, have presented Turkey with major dilemmas regarding the fulfillment of its officially stated policy objectives.

The eruption of ethnic and secessionist conflicts in Georgia, Nagorno-Karabakh, and Chechnya during the early 1990s raised Turkish concerns about their impact on regional stability and energy security. There was growing, and realistic, apprehension about the possibility of instability spilling over into Turkey, since the ethnic fighting took place close to Turkish borders and involved Turkic and/or Muslim peoples with whom Turkey had historic ties. In addition, there were sizeable numbers of Abkhazians, Azeris, and Chechens in Turkey who sympathized with their ethnic kindred across Turkey's borders in the Caucasus. The facts that the Azeris were pitted against the Armenians and the Chechens sought independence from the Russians were other sources of concern. Turkey has had historically difficult relations with both Armenia and Russia. The conflicts in the post–Cold War era threatened to exacerbate tensions with both states, especially in view of the growing military ties between Moscow and Yerevan. Turkish officials also worried that these conflicts would lead to increased regional polarization between Christians and Muslims and force Turkey to be part of an unwanted Huntingtonian "clash of civilizations" scenario.

Political instability and secessionist warfare in the Caucasus posed other security problems for Turkey. It was feared that separatist movements in Abkhazia, South Ossetia, and Chechnya could have a demonstration effect on the Kurdish problem in Turkey and undermine the country's territorial integrity. Moreover, Ankara was concerned about Moscow's use of overt and covert measures in the conflicts in Georgia and Nagorno-Karabakh to increase its strategic presence in the region. Turkish officials worried that the involvement of neighboring states in these ethnically based violent struggles for territory could lead to new alliances between regional states and neighboring powers, which in turn could increase the potential for the escalation and widening of these conflicts. There was also apprehension that ethnic strife and political instability could undermine Turkey's energy imports from the Caspian region and the construction of the Baku-Ceyhan pipeline.

Turkey's perspective on instability and conflict in the Caucasus and its desire for regional peace and stability reflected a perception that was widely held by Turkish foreign and defense policy specialists regarding the worsening of the country's strategic position in the 1990s.[21] According to this perception, despite the end of the Cold War rivalries and the demise of the Soviet threat, Turkey found itself in a far more dangerous new regional security environment and faced a number of major crisis points close to its borders in the Middle East, the Caucasus, and the Balkans. Moreover, the end of the Cold War intensified Greek-Turkish feuding in the Aegean and on Cyprus and heightened Turkish perceptions of the threat posed by Greece to Turkey's national security. At home, the Kurdish problem took a more violent turn in the 1990s and Turkey had to deploy more than one-third of its military force to contain the campaign of violence and terrorism that was spearheaded by the PKK. To make matters worse, the growing strains in Turkey's relations with its European allies in NATO meant that Ankara could not necessarily count on western support to deal with its growing list of security problems.

Under these circumstances, Turkey could ill-afford to become entangled in the ethnic and secessionist conflicts that erupted in the former Soviet Union. Possibly the most important test case regarding Turkey's potential and willingness to intervene in regional conflicts was the one between the Azeris and the Armenians over Nagorno-Karabakh. Turkey viewed Azerbaijan strategically as the most important state in the region due to its energy resources and because of its geopolitical significance for Russia and Iran. Other factors also added to the importance of Azerbaijan for Turkey. Of the five Turkic states in the South Caucasus and Central Asia, Turkey had the closest historic, ethnic, and linguistic ties with Azerbaijan.[22] By becoming the first country to recognize Azerbaijan's independence in 1991, Turkey reaffirmed the special relationship between the Azeris and the Turks. The presence of a large Azeri community in Turkey gave additional strength to the public sympathy and support for the Azeris in their conflict with the Armenians. Furthermore, under the presidency of Abulfaz Elchibey during 1992 and 1993, Azerbaijan pursued an openly pro-Turkish policy and relations between Baku and Ankara rapidly drew closer. In sharp contrast, Turkey's relations with Armenia have been characterized by enmity and hostility since the final days of the Ottoman Empire. Although Ankara and Erevan appeared to be interested in the normalization of relations following Armenia's independence, the eruption of the Nagorno-Karabakh conflict proved to be a major obstacle to closer ties between the two countries.

TURKEY AND THE NAGORNO-KARABAKH CONFLICT

Despite these conditions that seemed to favor Turkey's direct involvement on the side of the Azeris, in general, Ankara has chosen to stay out of the Nagorno-Karabakh conflict and refrained from providing troops and large-scale military support to Azerbaijan.[23] The growing domestic pressures on the Turkish government to intervene militarily led to strong official protests to Armenia that were accompanied by troop movements in eastern Turkey close to Armenia's border. Turkey sent a number of retired officers to Azerbaijan to train the Azeri troops, provided a limited amount of military supplies through clandestine channels, and offered humanitarian aid to Azeri refugees following the occupation of nearly one-fifth of Azerbaijan by Armenians. Ankara also imposed an embargo on Armenia in 1993 and became the major supporter of the Azeri view of the conflict in international diplomatic and political circles. But Turkey chose not to become militarily involved in the fighting between the Armenians and the Azeris. Clearly, it did not wish to become embroiled in ethnic warfare in the Caucasus when it faced several other major external and internal threats to its national security.

Turkish policy also took into account the potential problems that intervention in the Nagorno-Karabakh dispute could create in its relations with both Russia and the West. The strong Russian response to Turkish President Turgut Ozal's announcement in 1993 that Turkey might use military force to aid the Azeris underscored the possibility of a major conflict with Russia.[24] As a key player in national security policy, the Turkish military had no desire to risk a war with Russia in the Caucasus. In addition, direct military involvement carried the risk of undermining Turkey's relations with the United States and other NATO allies. Given the powerful Armenian lobby in Washington that successfully managed to craft a pro-Armenian position for the United States on the conflict, Turkey was concerned that any move to provide military support to the Azeri side could lead to new problems in its bilateral military and political relations with the United States. Similar problems could arise with European countries, and especially with France, which also has a sizeable Armenian diaspora community.

Turkey's interest in Caspian gas and oil production and the construction of the Baku-Ceyhan pipeline call for a peaceful resolution of the conflict between the Armenians and the Azeris rather than its escalation and internationalization.[25] Since the outbreak of the hostilities over Nagorno-Karabakh, Turkey has sought to play a role in the mediation efforts between the two sides. Russian and Armenian opposition to unilateral Turkish mediation quickly lowered Ankara's expectations concerning its role as a mediator and led to its support for multilateral peace initiatives. In 1992, Turkey became a member of the Minsk Group of countries that was formed to find a peaceful resolution to the Nagorno-Karabakh dispute under the auspices of the Organization on Security and Cooperation in Europe. The Minsk Group has been searching, so far unsuccessfully, for ways of establishing a comprehensive peace settlement that would satisfy both parties. Turkey is clearly interested in the settlement of the dispute and has tried to make good use of its close relations with Azerbaijan to move the Aliyev government toward a compromise solution. But both sides to the conflict appear to be unwilling to break the current stalemate, and external powers such as Turkey have discovered that they can only exercise limited influence regarding a peaceful resolution of the Nagorno-Karabakh problem.[26]

TURKISH POSITIONS ON CHECHNYA AND ABKHAZIA

Turkey's approach to the secessionist conflicts in Georgia and Chechnya reflected similar concerns. Apprehensive about their potential impact on

Russo-Turkish relations and its own Kurdish problem, Ankara declined to respond to the appeals of the Abkhazians and the Chechens for military support. In each case, however, there were reports of "volunteers" from Turkey who joined the fighting in Georgia and Chechnya. Their ranks were made up largely of ethnic Abkhazians and Chechens living in Turkey. In the case of the Chechnya conflict, militants from Turkey's far-right and radical Islamist groups, some of them veterans of the war in Afghanistan, also reportedly participated in the fighting.[27] While Turkey adopted a neutral stand in the conflicts that raged in Georgia in the early part of the 1990s, it came out strongly in support of the Chechens in their struggle against Moscow. Turkey's political support to the Chechens, the presence of volunteers from Turkey in the fighting, and the smuggling of armaments and weapons from Turkey prompted Moscow to launch strong protests to Ankara. Russia also expressed its anger by threatening to play the "Kurdish card" against Turkey by supporting the PKK. In implementing these measures, the Russian government allowed the PKK to open an office in Moscow and to organize meetings in the Russian Parliament building. Coming after several decades of covert support by the former Soviet Union to militant Kurdish groups in Turkey, Russia's policy toward the PKK in the 1990s renewed Turkish suspicions about Moscow's intentions and became a new irritant in relations between the two countries.

Turkey's response to ethnic and separatist conflicts in the Caucasus during the 1990s suggests that Ankara is likely to pursue a cautious approach and refrain from direct military involvement in regional conflicts in the near future. A unilateral military intervention on behalf of the Azeris or other Turkic and Muslim communities carries high risks for Turkey at a time when the country is already burdened with external and domestic national security problems. Ankara continues to maintain close military ties with Baku that include the training of several hundred Azeri officers and cadets in Turkey. In case of renewed armed hostilities between the Armenians and the Azeris, Turkey could provide limited amounts of military aid to the Azeris. However, it is unlikely that Turkey would send soldiers and large-scale armaments to Nagorno-Karabakh. While new activism in Turkish foreign policy led Ankara to flex its military muscle against Syria and the Greek Cypriot administration during the recent crises concerning the expulsion of PKK leader Abdullah Ocalan from Damascus and the delivery of S-300 missiles from Russia to Cyprus, the situation in the Caucasus is quite different. While Turkey might be willing militarily to take on Syria, or possibly even Greece, to protect its perceived national interests, it would be extremely reluctant to engage in a military conflict with Russia, a nuclear power with a much larger military force. Turkish military intervention on the side of the Azeris would almost certainly bring the Russians and possibly the Iranians into open conflict with Turkey. Under the present circumstances, no government in

Ankara can afford to embark on such a risky policy that could rapidly lead to the internationalization of a local conflict—with potentially adverse consequences for Turkey.

NEIGHBORING POWERS COMPETE FOR INFLUENCE IN THE CASPIAN REGION

Instability and ethnic conflicts in the Caucasus have contributed to increased geopolitical competition between Russia, Iran, and Turkey. Moscow is uncomfortable with Ankara's new activist foreign policy directed at Azerbaijan and the Central Asian Turkic republics, an attitude based not so much on fear of pan-Turkism as on concern over an increase of Turkey's regional influence after a long period of Turkish quiescence in the region. At the same time, Russia's efforts to reassert its dominance over the former Soviet Union's territories in the Caucasus have rekindled traditional Turkish perceptions about Russia. Rooted in the historical conflicts between the two countries, especially Russia's expansion into the Caucasus at the expense of the Ottoman Empire, these perceptions continue to shape Turkey's outlook on Russian policies in the Caspian region. There is a fairly widespread view that Russia seeks to limit Turkey's regional role while expanding its power and influence. Ankara's inability or unwillingness to support President Elchibey when he was overthrown by a Moscow-backed rebellion in June 1993 underscored Turkey's limited capability to compete effectively with Russia. Growing military cooperation between Armenia and Russia in the late 1990s, including the shipment of S-300 missile systems and advanced MiG-29 fighter jets to Armenia, has heightened Turkish security concerns. The military build-up of Armenia and the growing Russian military presence close to Turkish borders could lead to the expansion of Turkey's military ties with Azerbaijan.[28]

While competition and rivalry have characterized Russian and Turkish policies regarding geostrategic and security issues in the Caucasus, trade and commercial ties between the two countries have expanded at a rapid pace since the collapse of the Soviet Union, enabling Turkey to become one of Russia's major trading partners. Turkish construction companies have undertaken large-scale projects in the Russian Federation, and Turkey has become a major source of Russian imports of food items and consumer goods. Turkish construction firms and trading companies that have strong business ties in Russia have become an influential force in shaping Ankara's policies toward Moscow. The "Russian lobby" in Turkey has a strong stake in the improvement of political relations and the lessening of geopolitical rivalries between the two countries. The key question for the near future, therefore, is whether Turkey's policies toward Russia will be driven largely by geopolitical or economic and commercial concerns.

Turkey's relations with Iran have also felt the impact of regional conflicts and geopolitical rivalries. Each state has viewed the other's efforts to gain political and economic influence in the South Caucasus and Central Asia with considerable mistrust. In the post–Cold War era, Iran was concerned that "any lessening of Soviet influence in this area would be replaced by an increase in Turkish influence, and by extension in American influence."[29] In particular, the Iranian leadership was apprehensive about Turkey's efforts to forge close political, economic, and cultural ties with Azerbaijan. Tehran's concerns became especially pronounced during the presidency of Elchibey, who pursued an openly pro-Turkish and anti-Iranian policy. Iran was worried that the strengthening of pro-Turkish sentiments in Azerbaijan could alter the regional strategic balances to its disadvantage. Turkey, for its part, viewed the possibility of increased Iranian influence in Azerbaijan as an obstacle to its efforts to pursue its political and economic interests. Although the competition between Iran and Turkey over Azerbaijan has abated somewhat since the early 1990s, geopolitical concerns continue to shape each country's perceptions of the other's regional policies and behavior.[30]

Despite geopolitical rivalries and periodic crises in their political relations, Turkey and Iran have sought to expand their economic and commercial ties. As mentioned, Turkey views Iran as an important source for energy and a potentially important market for its exports. Similarly, Tehran sees its western neighbor as a profitable market for its natural gas and petroleum products and a relatively cheap source of imported manufactured goods and consumer products. As in the case of Turkey's relations with Russia, it remains to be seen if commonly shared economic and commercial interests, rather than political and ideological differences, will be the main determinants of Turkey's policies toward Iran in the future.

DOMESTIC STABILITY AND ENERGY SECURITY

Turkey's growing focus on Caspian energy development has taken place in the context of instability and uncertainty in domestic politics. The absence of strong, durable, and effective governments in Ankara from 1991 to 1999 has been the single most important cause of political uncertainty. Governmental instability has also undermined the formulation and implementation of policy. In terms of significant Caspian energy issues, rapid turnover of ministers and top bureaucrats in charge of energy policy, and bureaucratic infighting between agencies following each new government's rise to power, hindered policy continuity and consistency.[31]

The main reason for the rise and fall of weak and unstable coalition or minority governments has been increased political fragmentation in the Turkish party system. National elections in 1991 and 1995 failed to provide any one party with a majority of seats in the Parliament. These inconclusive election outcomes necessitated the formation of coalition or minority governments that could not provide effective and stable political leadership. The elections in 1999 also resulted in the formation of a coalition government. However, unlike its predecessors, the new three-party coalition government headed by Prime Minister Bulent Ecevit controls a strong majority in the Parliament.[32]

Secularists Versus Islamists

During the 1990s, governmental instability in Turkey was accompanied by increasing polarization between the secularists and the Islamists and the intensification of the Kurdish problem. While the issue of political Islam threatens Turkey's established secular constitutional order, the Kurdish question poses a serious challenge to the country's territorial integrity. It is difficult to predict how the ongoing conflict between the pro-secular and the Islamist forces will play out in the near future. The emergence of the Islamist Welfare Party as the largest parliamentary force in the 1995 elections came as a shock to the country's secularist establishment.[33] The decline in the electoral strength of the Islamists in the 1999 elections has not diminished the concerns of the pro-secular Turks regarding political Islam. Most secularists are convinced that the Islamists are using democratic processes simply as a means to achieve their goal of replacing the country's constitutional order with an authoritarian Islamic regime. The Turkish military, which has been a staunch supporter of secularist policies, played a major role in the downfall of the Islamist-led coalition government in 1997 and seems determined to prevent the Islamists from assuming power again. The Islamists have contributed to the intensification of the conflict through their fiery rhetoric against secularism and western-style parliamentary democracy during their long years in the opposition. When they were in power in 1996 and 1997, the Islamists toned down their political rhetoric and tried to pursue a relatively moderate course. Nevertheless, most secularists continue to believe that this apparent turn to moderation is only a political tactic designed to neutralize the opposition—especially military opposition—to the Islamists.

The polarization between the secularists and the Islamists has led to a number of political crises in recent years and has severely strained Turkey's fragile democracy. However, it has not been a source of political violence. Turkey's Islamists, with the exception of some minuscule radical fringe groups, have chosen to pursue peaceful strategies to attain their objectives. There is little evidence that they will change their approach in the near future. Consequently, the secularist versus Islamist conflict is not likely to significantly affect Turkey's Caspian energy and pipeline projects. If the Islamists manage to come to power again, they will be under the close scrutiny of the military, which maintains a strong interest in energy issues. It is also unlikely that an Islamist government would follow an activist policy in the Caucasus. When the Welfare Party's coalition government was in power, it sought to strengthen ties with other leading Muslim states, but virtually ignored Central Asia and the Caucasus. However, an Islamist government can be expected to seek increased political and economic cooperation with Iran, including on energy issues. The Islamists enthusiastically supported the 1996 Iranian-Turkish natural gas deal that was signed while an Islamist-led coalition government was in power in Ankara.

Continued Conflict with the Kurds

Unlike political Islam, Turkey's Kurdish problem has involved a great deal of political violence.[34] Since 1984, Turkey has sought to control the campaign of political violence and terrorism that was initiated by the PKK. During fifteen-plus years of fighting, more than 30,000 people—PKK militants, Turkish military and police, and civilian noncombatants—have lost their lives, and thousands of others have been wounded. The PKK, which began as a Marxist-Leninist organization that sought to establish an independent Kurdish state, has received substantial support from Syria. Other neighboring countries, such as Iran and Greece, have also backed Kurdish military efforts through overt and covert means. Since the mid-1990s, the Turkish military has managed to control the PKK's activities in southeastern Turkey by attacking PKK bases in northern Iraq, aerial bombing, and the evacuation of nearly three thousand hamlets and villages in the southeast to deprive the PKK of sustenance and support. The capture of Abdullah Ocalan in early 1999 represents another important blow at the PKK, since he had almost single-handedly been responsible for the organization's development, methods, and strategies.

Despite the military successes on the ground against the PKK, Turkey continues to face a critical problem regarding the accommodation of demands voiced by many Kurds for greater cultural and political rights. During the past decade, there have been some important changes in the official outlook on the Kurds: the 1983 language law that banned the use of Kurdish was repealed; the need to recognize a distinctive Kurdish identity was emphasized by high-ranking government officials; and pro-Kurdish political parties have been permitted to compete in elections. However, these measures have failed to satisfy Kurdish demands for education in their own language, radio and TV stations that broadcast in Kurdish, and a more decentralized administrative system for Turkey's predominantly Kurdish provinces in the southeast. Possible reforms have been stymied due primarily to the priority that Turkey has attached to militarily suppressing the PKK. So far, other policy options that might lead to a settlement of the dispute have not entered the policy-making process. As one observer has noted: "For even the most liberal Turkish policy makers, the struggle with the PKK presents a problem. They could make concessions and possibly whet the separatist appetite, or fight on, conceding nothing, and possibly fuel the resentment that fosters separatism."[35]

Under the circumstances, it would not be realistic to expect an early resolution of the Kurdish problem in Turkey. However, Ocalan's capture and trial have significantly changed the dynamics of the conflict. Upon Ocalan's orders, the PKK declared an end to its campaign of political violence and began moving its militants out of Turkey. As a result, there has been a sharp drop in clashes between the security forces and the PKK since the latter part of 1999. Although it has suffered major losses against the authorities, the PKK is likely to be around in the near future, thanks to the support it receives from a segment of the Kurdish population in Turkey and among the Kurdish migrants in Europe and some of Turkey's neighbors. Moreover, as an organization that is deeply involved in the trafficking of drugs and illegal migrants from the Middle East to Europe, the PKK will continue to command the necessary financial resources to remain active in Turkey and abroad.

In the past, the PKK has repeatedly warned that it would try to sabotage Caspian energy pipelines in eastern Turkey to hurt the country economically and to undermine its strategic importance as a transit state for the transport of Caspian oil and gas to western markets. Turkish officials have dismissed these claims and declared that Baku-Ceyhan and other pipeline projects will be well protected. The PKK's decision to cease its activism makes it highly unlikely that the proposed new pipelines would be the targets of a sabotage attempt by the Kurdish militants in the near future. Even if the PKK changes its strategy and resumes its activities, however, the "vulnerability of oil pipelines [should not] be exaggerated. Sabotaged or damaged pipelines can be repaired, pumping stations can be shut off to compartmentalize damage and curb the loss of oil and environmental damage, and oil can be made to flow again without prolonged interruptions."[36]

IMPLICATIONS FOR U.S. POLICY

Cooperation in Caspian energy development has been one of the main pillars of U.S. policy toward Turkey during the 1990s. The strong U.S. support for the Baku-Ceyhan project has been the cornerstone of U.S.-Turkish cooperation in Caspian energy issues. Washington's preference for Baku-Ceyhan as the principal transit route for Azerbaijani oil within a multiple-pipelines scheme reflects the Clinton administration's desire to strengthen Turkey's ties with the West and to curb Russian and Iranian economic and geostrategic interests in the Caspian region.[37] Turkey has acknowledged the important role that the United States has played in the promotion of the Baku-Ceyhan project. Cooperation on this issue has been an important factor in the maintenance of strong bilateral political and military ties between the two countries despite the changes that were brought about by the end of the Cold War. Washington continues to see Turkey as a key regional state that is strategically important for U.S. national interests. Turkey regards the United States as its principal ally, and Turkish officials have assigned even greater importance to the bilateral relationship due to the strains in Turkey's relationship with Europe.

One of the potential problems for U.S. policy concerns the fallout that would inevitably result if the Baku-Ceyhan project fails to materialize. Washington's support for the project has increased expectations that Turkey will emerge as the main transit state for Azerbaijani oil in the first decade of the twenty-first century. As noted earlier, Baku-Ceyhan has become a matter of national pride and a symbol of Turkey's role and influence in the Caucasus. While Washington has staked much prestige on Baku-Ceyhan, it has not been able to convince the oil companies to commit to this route. If Baku-Ceyhan does not materialize, it will be a major disappointment for Turkey and lessen Washington's credibility in Ankara. This could, in turn, undermine the strength of the bilateral relationship as well as U.S. efforts to anchor Turkey firmly in the West.

Another problem in U.S. policy concerns the role of the oil companies in Caspian pipeline development. Until the late 1990s, Washington focused most of its effort on political and diplomatic measures and gave relatively less importance to the economic interests of the western oil companies, who have been concerned about the lack of financial incentives and the commercial viability of the Baku-Ceyhan project. Until recently, Turkish policymakers appeared to believe that political support from Washington and Baku would be sufficient to overcome the opposition of the oil companies. With its recent decision to provide guarantees for construction costs exceeding $2.7 billion, Turkey has belatedly recognized the need to address these commercial interests. Clearly, the future of Baku-Ceyhan will depend, among other factors, on the ability and willingness of Ankara and Washington to show greater sensitivity to the economic concerns of the oil companies.

In terms of the geopolitics of the Caspian region, the United States and Turkey share similar views regarding the possible expansion of Russian and Iranian political and economic influence. Both Washington and Ankara are apprehensive about the reestablishment of Russian dominance in the South Caucasus, although Turkish officials have been far more concerned about Moscow's policies than have their American counterparts. Clearly, Ankara would welcome a more assertive U.S. policy concerning Moscow's efforts to implement its "near abroad" doctrine through strong-armed tactics in the region and the growing military cooperation between Russia and Armenia, both of which threaten to undermine regional strategic balances and increase the likelihood of conflicts.

For Turkey, one of the major problems with U.S. policy concerns Washington's differential approach to Armenia and Azerbaijan since the end of the

Cold War. The Armenian diaspora community in the United States, through its powerful lobby in Washington, managed to move the Bush and Clinton administrations toward a pro-Armenian position in the conflict over Nagorno-Karabakh. While Armenia has been the second highest recipient of U.S. foreign aid on a per capita basis after Israel, Azerbaijan has been excluded from U.S. assistance by Section 907 of the Freedom Support Act. Recent modifications to Section 907 and the exemption of a number of restrictions that were originally imposed on Azerbaijan are signs of perceptible changes in U.S. policy toward Azerbaijan. Nevertheless, there is widespread belief among the Turkish public and officials that Washington continues to reward the Armenians and punish the Azeris, despite the fact that more than 20 percent of the territory of Azerbaijan has been occupied by Armenians since the early-1990s and nearly one million Azeris have been displaced from their homes and forced to live as refugees. Turkey has voiced its opposition to the imbalance in U.S. policies and has repeatedly called on Washington to adopt a more even-handed approach. A change in U.S. policy, one that recognizes the interests and concerns of both Azerbaijan and Armenia, would be welcomed in Ankara and remove an irritant from U.S.-Turkish relations.

NOTES

1. For a perceptive analysis, see Alan Makovsky, "The New Activism in Turkish Foreign Policy," *SAIS Review* 29, no. 1 (Winter–Spring 1999): 92–113. See also Andrew Mango, *Turkey: The Challenge of a New Role* (Westport, Conn: Praeger, 1994); and Kemal Kirisci, "New Patterns of Turkish Foreign Policy Behavior," in *Turkey: Political, Social and Economic Challenges in the 1990s,* ed. Cigdem Balim, et al. (Leiden: E.J. Brill, 1995), 1–21.

2. Sabri Sayari, "Turkey, the Caucasus, and Central Asia," in *The New Geopolitics of Central Asia and Its Borderlands,* ed. Ali Banuazizi and Myron Weiner (Bloomington: Indiana University Press, 1994), 175–196.

3. See Alvin Z. Rubinstein and Oles M. Smolansky, eds., *Regional Power Rivalries in the New Eurasia: Russia, Turkey, and Iran* (Armonk, N.Y.: M. E. Sharpe, 1995).

4. See Victor A. Nadein-Raevsky, "Some Opinion on the Turkic Factor," in *Russia: The Mediterranean and the Black Sea Region,* ed. Nicolai A. Kovalsky (Moscow: Russian Academy of Sciences, 1996), 252–266.

5. See Idil Tuncer, "Rusya Federasyonunun Yeni Guvenlik Doktrini: Yakin Cevre ve Turkiye (The Russian Federation's New Security Doctrine: Near Abroad and Turkey)," in *En Uzun On Yil: Turkiye'nin Ulusal Guvenlik ve Dis Politika Gundeminde Doksanli Yillar (The Longest Decade: Turkey's National Security and Foreign Policy Agenda During the 1990s),* ed. Gencer Ozcan and Sule Kut (Istanbul: Boyut Kitaplari, 1998), 445–471.

6. The opening of the Baku-Supsa oil pipeline from Azerbaijan to Georgia in April 1999 effectively ended Russia's near-monopoly over the transportation of oil from Azerbaijan.

7. See Justin McCarthy, *Death and Exile: The Ethnic Cleansing of Ottoman Muslims, 1821–1922* (Princeton, N.J.: Darwin Press, 1995), 179–253.

8. William Hale, *The Political and Economic Development of Modern Turkey* (New York: St. Martin's Press, 1981), 203–204.

9. Hale, *The Political and Economic Development of Modern Turkey*, 203.

10. See Philip Robins, *Turkey and the Middle East* (New York: Council on Foreign Relations, 1991), 74–113.

11. Sabri Sayari, Simon Blakey, and Shakari Sribivasan, "New Bridges to Turkey's Hydrocarbon Future," *Cambridge Energy Research Associates Private Report* (Cambridge, Mass.: Cambridge Energy Research Associates, April 1998).

12. International Energy Agency (IEA), *Energy Policies of IEA Countries: Turkey 1997 Review* (Paris: IEA, 1997), 49.

13. Sayari, et al., "New Bridges to Turkey's Hydrocarbon Future," 1.

14. Estimates for the construction of Baku-Ceyhan have been a source of controversy since the early 1990s. In an April 1999 meeting with oil company officials in Istanbul, the Turkish government announced that it would guarantee price costs over $2.4 billion. Two months later, after a new government came to power in Ankara following parliamentary elections, Turkey announced that it was raising the price cap to $2.7 billion.

15. Stephen Kinzer, "Turks Fear an Oil Disaster as the Bosporus Gets Busier," *New York Times*, January 11, 1998, A3.

16. The sensationalist press coverage was reflected in the headlines of Istanbul dailies. For example, when the United States announced its support for Baku-Ceyhan, the headline in a major daily stated "Petrolde Zafer Turkiye'nin" (Victory in Petroleum Belongs to Turkey). See *Milliyet,* October 10, 1995.

17. Rajan Menon, "Treacherous Terrain: The Political and Security Dimensions of Energy Development in the Caspian Sea Zone," *NBR Analysis* (The National Bureau of Asian Research) 9, no. 1 (February 1998): 28.

18. Stephen Kinzer, "Oil Pipelines from Caspian Lack Money from Backers," *New York Times*, November 28, 1998, A1.

19. "Trans-Caspian Gas Pipe Plans Advance," *Caspian Investor* 2, no. 3 (November–December 1998): 7. For information on the politics and economics of the Turkmen gas projects, see chapter 6, "Turkmenistan's Energy: A Source of Wealth or Instability?" by Nancy Lubin, in this volume. Despite strong support from Washington, the trans-Caspian energy corridor project does not have the necessary financing. In addition, Turkmenistan and Azerbaijan continue to feud over a division of the Caspian oil fields. Similarly, there has been no progress in the Blue Stream proposal due to shortage of funds for a project that is expected to cost approximately $1.2 billion. Faced with uncertainties about the future of these two projects, Turkey has proceeded with the construction of the pipeline from Ankara to Erzurum, which will eventually be linked with a pipeline from Iran. At present, the Iran-Turkish gas pipeline and purchase project appears to have the best chance for early completion.

20. On the evolution of Turkey's political relations with the Caspian region, see Suha Bolukbasi, "Turkey's Baku-Centered Transcaucasia Policy: Has it Failed?" *Middle East Journal* 51, no. 1 (Winter 1997): 81–94; Shireen Hunter, *The Transcaucasus in Transition* (Washington, D.C.: Center for Strategic and International Studies, 1994), 161–170; and William Hale, "Turkey and Transcaucasia," in *Central Asia Meets the Middle East,* ed. David Menashri (London: Frank Cass, 1998), 150–167.

21. High ranking Turkish military officers have emphasized the country's worsening security problems as a result of increasing external threats at the annual meetings of the American-Turkish Council in Washington, D.C. For a summary, see "p. 389 *Nokta,* June 20–26, 1999, 44–46.

22. Sayari, "Turkey, the Caucasus, and Central Asia," 178–179.

23. For a perceptive analysis, see Svante E. Cornell, "Turkey and the Conflict in Nagorno-Karabakh: A Delicate Balance," *Middle Eastern Studies* 34, no. 1 (January 1998): 51–72.

24. Cornell, "Turkey and the Conflict in Nagorno-Karabakh: A Delicate Balance," 65.

25. As a ranking Turkish Foreign Ministry official put it "[if] the Azerbaijani oil is to be brought down to the Mediterranean . . . the security of the pipeline should be ensured. This can not be achieved while Azerbaijan is at war." See the comments of Ambassador Halil Akinci in *Turkey at the Threshold of 21st Century,* ed. Mustafa Aydin (Ankara: International Relations Foundation, 1998), 135.

26. Patricia Carley, *Nagorno-Karabakh: Searching for a Solution: United States Institute of Peace Roundtable Report* (Washington, D.C.: United States Institute of Peace, December 1998).

27. For a report on the Turkish Islamists who participated in the Chechen war, see "Cecenistan'da Denedik (We Tried it in Chechnya)," *Nokta,* April 30–May 6, 1995: 8–13.

28. In early 1999, Azerbaijan proposed that the United States and Turkey should build bases in Azerbaijan to counter the growing Russian presence in Armenia. Neither country, however, has taken up this offer. See Stephen Kinzer, "Azerbaijan Asks the US to Establish Military Base," *New York Times,* January 31, 1999, A5.

29. Fred Halliday, "Condemned to React, Unable to Influence: Iran and Transcaucasia," in *Transcaucasian Borders,* ed. John F. R. Wright, et al. (New York: St. Martin's Press, 1996), 82.

30. Clearly, other political and ideological issues, particularly Ankara's concern about Iran's support for Turkey's Islamist groups and the PKK, plays an important role in Turkish perceptions about Iran.

31. For an account of how government instability and bureaucratic infighting have hampered Turkey's pipeline policies, see Lale Sariibrahimoglu, *Kurt Kapaninda Kisir Siyaset: Gizli Belgelerle Boru Hatti Bozgunu (Unproductive Politics in the Wolf's Trap: Pipeline Defeat According to Secret Documents)* (Ankara: Imge Yayinlari, 1997).

32. The new coalition government includes the far-right Nationalist Action Party (NAP) that has long been a strong supporter of closer ties between Turkey and the Central Asian Turkic states. NAP has been particularly active in Azerbaijan in the 1990s. Its participation in the government might lead to a hardening of Turkey's position vis-à-vis Russia in the Caucasus.

33. See Sabri Sayari, "Turkey's Islamist Challenge," *Middle East Quarterly* 3, no. 3 (September 1996): 35–44.

34. On the Kurdish issue, see Kemal Kirisci and Gareth M. Winrow, *The Kurdish Question and Turkey* (London: Frank Cass, 1997); and Henri J. Barkey and Graham E. Fuller, *Turkey's Kurdish Question* (Lanham, Md.: Rowman and Littlefield, 1998).

35. Alan O. Makovsky, "Turkey," in *The Pivotal States,* ed. Robert Chase, Emily Hill, and Paul Kennedy (New York: W.W. Norton, 1999), 110.

36. Menon, "Treacherous Terrain," 27.

37. On the relationship among the United States, Iran, and Russia in the context of the Caspian region, see chapter 8, "U.S.–Iranian Relations: Competition or Cooperation in the Caspian Sea Basin," by Geoffrey Kemp.

Index

About the Contributors

Robert Ebel is director of the Energy and National Security Program at the Center for Strategic and International Studies. Formerly vice president for international affairs at Enserch Corporation, Mr. Ebel advised the corporation on global issues. Mr. Ebel traveled to the Soviet Union in 1960 as a member of the first U.S. petroleum industry delegation, and in 1971 with the first group of Americans to visit the oil fields in Western Siberia. Mr. Ebel is author of *Energy Choices in the Near Abroad: The Haves and Have-nots Face the Future and Energy Choices in Russia.*

Dru C. Gladney is professor of Asian studies and anthropology at the University of Hawaii (on leave) and a special researcher on Central Asia and China at the Asia-Pacific Center for Security Studies in Honolulu. His recent books include: *Ethnic Identity in China* (1998); *Making Majorities: Constituting the Nation in Japan, China, Malaysia, Indonesia, Turkey, and the United States* (ed., 1998), and *Muslim Chinese: Ethnic Nationalism in the People's Republic of China* (1992, 2nd edition).

David I. Hoffman is senior associate at Cambridge Energy Research Associates and a doctoral candidate in political science at the University of California-Berkeley. His research focuses on the effects of energy-sector developments on state building in Azerbaijan and Kazakhstan.

Shireen T. Hunter is the director of Islamic Studies at the Center for Strategic and International Studies. She previously served as director of the Mediterranean Studies program with the Centre for European Policy Studies in Brussels. Dr. Hunter is author of *The Future of Islam and the West: Clash of Civilizations or Peaceful Coexistence?* (1998), *Central Asia Since Independence* (1996), and *The Transcaucasus in Transition: Nation-Building and Conflict* (1994). Her articles have appeared in leading journals such as *Foreign Affairs, Foreign Policy, Current History, Middle East Journal, Security Dialogue, the International Spectator,* and *SAIS Review.*

Terry Lynn Karl is associate professor of political science and director of the Latin American Studies Center at Stanford University. Dr. Karl specializes in comparative politics, the political economy of development, U.S. foreign policy, comparative democratization, and conflict resolution. Dr. Karl is author of *The Paradox of Plenty: Oil Booms and Petrostates* (1997) and *Limits to Competition.*

Geoffrey Kemp is director of Regional Strategic Programs at The Nixon Center for Peace and Freedom. He served in the White House during the first Reagan Administration as Special Assistant to the President for National Security Affairs and Senior Director for Near East and South Asian Affairs on the National Security Council staff. Dr. Kemp is the author of *Strategic Geography and the Changing Middle East* (1997); *Point of No Return: The Deadly Struggle for Middle East Peace* (1997); and *Forever Enemies? American Policy and the Islamic Republic of Iran* (April 1994).

Nancy Lubin is president of JNA Associates, Inc.—a research and consulting firm focusing on the new states of the former Soviet Union, especially Central Asia—and Director of the Council on Foreign Relations Project on the Ferghana Valley. She has lived, worked, and travelled throughout Central Asia and the Caucasus for twenty five years and is the author of books, Congressional testimony, and a wide range of scholarly and popular articles.

Pauline Jones Luong is assistant professor of political science at Yale University. Her primary research interests include: institutional origin and change; ethnic identity, cooperation and conflict; and the political economy of development. She has published articles on Central Asian politics and society in *International Negotiation, Journal of International Affairs,* and *Central Asian Monitor.* She is working on a new book project, that investigates the short- and long-term impact of natural resource endowments on state formation in developing countries, focusing on the energy-rich Soviet successor states.

Michael Mandelbaum is Christian A. Herter Professor and director of American Foreign Policy at the School of Advanced International Studies, Johns Hopkins University, and director of the Council on Foreign Relations Project on East-West Relations. Dr. Mandelbaum is an internationally recognized authority on world politics, U.S. foreign policy, and the international relations of the post-Soviet states. He has written and edited numerous publications on foreign policy, including: *The New Russian Foreign Policy* (1998); *The Dawn of Peace in Europe* (1996); and *Central Asia and the World* (1994).

Rajan Menon is Monroe J. Rathbone Professor of International Relations at Lehigh University and director of the Eurasia Policy Studies program at The National Bureau of Asian Research. In addition, Dr. Menon is Academic Fellow and advisor for the International Peace and Security Program at the Carnegie Corporation of New York. Professor Menon is author and editor of numerous publications on the international relations and security in Russia and the other newly independent states of the former Soviet Union, including: "The Limits of Neo-Realism: The Case of Conflict and Conflict Resolution in Central Asia," *Review of International Studies* (February 1999), *Russia, The South Caucasus, and Central Asia: The Emerging 21st Century Security Environment* (1999); and "Treacherous Terrain: The Political and Security Dimensions of Energy Development in the Caspian Sea Zone," *NBR Analysis* (1998).

Martha Brill Olcott is professor of political science at Colgate University, Senior Associate at the Carnegie Endowment for International Peace, and Codirector of the Carnegie Moscow Center's project on Ethnicity and Politics in the Former Soviet Union. She specializes in Central Asian affairs and inter-ethnic relations in the Soviet successor states. Dr. Olcott is the author of *Central Asia's New States: Independence, Foreign Policy, and Regional Security* (1996); "Islam and Fundamentalism in Independent Central Asia," in *Muslim Eurasia: Conflicting Legacies* (1995); and "Sovereignty and the 'Near Abroad'," *Orbis* (Summer 1995).

Peter Rutland is associate professor of government at Wesleyan University, where he specializes in the politics and economic policies in the former Soviet Union. He has published numerous books and articles on the political and economic transitions of the former Soviet republics, including: "Lost Opportunities: Energy and Politics in Russia," *NBR Analysis* (1997); "Russia's Broken 'Wheel of Ideologies'," *Transitions* (June 1997); "Russia's Unsteady Entry into the Global Economy," *Current History* (October 1996); and "Democracy in Armenia," *Europe/Asia Studies* (1994).

Sabri Sayari is executive director of the Institute of Turkish Studies and professor of international relations at Georgetown University. He is also chairperson of Turkish Area Studies at the State Department's Foreign Service Institute and a trustee of the Azerbaijan-United States Chamber of Commerce. Dr. Sayari has published extensively on Turkey's domestic and foreign policy, and on issues related to political development, democratization, and international political economy.

The National Bureau of Asian Research

The National Bureau of Asian Research (NBR) is a nonprofit, nonpartisan organization devoted to bridging the policy, academic, and business communities with advanced, policy-relevant research on issues confronting Asia and the former Soviet Union. Through publications, conferences, television programs, and other projects, NBR serves as an international clearinghouse on a wide range of issues, from trade and investment to national security. NBR does not take policy positions, but rather sponsors studies by the world's leading specialists to promote the development of effective and far-sighted policy. Recent projects have focused, for example, on China's most-favored-nation trade status, the protection of intellectual property rights in China, energy developments in Russia and the Caspian region, the evolving security environment in Southeast Asia, and the Asia-Pacific Economic Cooperation (APEC) forum.

NBR's research agenda is developed and guided by a bipartisan board of advisors, composed of individuals drawn from academia, business, and government, including 69 U.S. senators and representatives. Its operations are overseen by a distinguished national board of directors. NBR was founded in 1989 with a major grant from the Henry M. Jackson Foundation.